About Island Press

Since 1984, the nonprofit Island Press has been stimulating, shaping, and communicating the ideas that are essential for solving environmental problems worldwide. With more than 800 titles in print and some 40 new releases each year, we are the nation's leading publisher on environmental issues. We identify innovative thinkers and emerging trends in the environmental field. We work with world-renowned experts and authors to develop cross-disciplinary solutions to environmental challenges.

Island Press designs and implements coordinated book publication campaigns in order to communicate our critical messages in print, in person, and online using the latest technologies, programs, and the media. Our goal: to reach targeted audiences—scientists, policymakers, environmental advocates, the media, and concerned citizens—who can and will take action to protect the plants and animals that enrich our world, the ecosystems we need to survive, the water we drink, and the air we breathe.

Island Press gratefully acknowledges the support of its work by the Agua Fund, Inc., Annenberg Foundation, The Christensen Fund, The Nathan Cummings Foundation, The Geraldine R. Dodge Foundation, Doris Duke Charitable Foundation, The Educational Foundation of America, Betsy and Jesse Fink Foundation, The William and Flora Hewlett Foundation, The Kendeda Fund, The Forrest and Frances Lattner Foundation, The Andrew W. Mellon Foundation, The Curtis and Edith Munson Foundation, Oak Foundation, The Overbrook Foundation, the David and Lucile Packard Foundation, The Summit Fund of Washington, Trust for Architectural Easements, Wallace Global Fund, The Winslow Foundation, and other generous donors.

A Critique of Silviculture

A CRITIQUE OF SILVICULTURE

Managing for Complexity

Klaus J. Puettmann
K. David Coates
Christian Messier

Washington · Covelo · London

Library of Congress Cataloging-in-Publication Data

Puettmann, Klaus J.
A critique of silviculture : managing for complexity / Klaus J. Puettmann, K. David Coates, Christian Messier.
pp. cm.
Includes bibliographical references and index.
ISBN-13: 978-1-59726-145-6 (cloth : alk. paper)
ISBN-10: 1-59726-145-9 (cloth : alk. paper)
ISBN-13: 978-1-59726-146-3 (pbk. : alk. paper)
ISBN-10: 1-59726-146-7 (pbk. : alk. paper)
1. Forests and forestry—North America. 2. Forest management—North America. 3. Forest ecology—North America. 4. Forest biodiversity conservation—North America. 5. Logging—North America. I. Coates, K. Dave. II. Messier, Christian C. III. Title.

SD391.P97 2008
634.9′50973—dc22

2008010304

Printed on recycled, acid-free paper

Manufactured in the United States of America

10 9 8 7 6 5 4 3 2 1

Contents

Preface

"Life used to be simple." Memory may play funny games with us, but most would agree that our personal and professional lives have become more complicated. Silviculture and the management of forested ecosystems are certainly no exception. For most of the twentieth century, silviculture professionals were respected and their decisions regarding management practices were rarely questioned or challenged by the general public. Students entering silviculture and other forestry programs had a clear vision of their future. Silviculturists were successful at achieving clearly defined management objectives that usually emphasized efficient wood production. Silviculture had developed into a solid scientific discipline and was considered a central part of forestry research, teaching, and management programs.

Today, methods and techniques employed by silviculturists to manage forests are frequently challenged. Educational programs in forestry are struggling to maintain sufficient enrollment, staffing in public management organizations is constantly reduced, and job security in the forest industry is a thing of the past. The state of the profession is gloomy and the public's romantic view of silviculture and forestry has been lost. How could such a long-term trend of success in the silvicultural management of forests reverse itself in such a short time period? Such dramatic changes can be explained only by a combination of factors. Silviculture specifically, and forestry in general, did not keep up with the rapid changes in people's expectations and the increased complexity of modern twenty-first-century life.

It is very apparent to us that silviculture—and, more broadly, forest management—now needs to go through unprecedented changes and focus on different values. There is increased concern about the disappearance of old-growth and primary forests all over the world and the role of managed forests in the maintenance of biodiversity, carbon budgets, and the provision of many other ecosystem services. At the same time, we are gaining a better understanding of the multitude of environmental services that natural and managed forests provide. Silviculturists must address these developments and respond to the rapidity of changes in expectation and global paradigm shifts in how forests are viewed.

The discipline of silviculture appears to be at a crossroads. Silviculturists are being challenged to develop practices that sustain the full function and dynamics of forested ecosystems and maintain ecosystem diversity and resilience while still providing needed wood products. This book provides a critical re-evaluation of basic silvicultural assumptions and approaches in light of the new demands on silviculture in the twenty-first century. We then propose that silviculture requires a new conceptual framework to effectively address these issues. The new framework should come from ecology and complex systems science. We present our ideas of how silviculture can benefit from an improved understanding of ecological complexity and complex adaptive systems, especially ways to incorporate aspects of variability and uncertainty into management decisions.

—Klaus Puettmann, David Coates, and Christian Messier
January 2008

Acknowledgments

This book is the product of many discussions among the authors. We owe the inspiration for this book to the increasing criticisms that silviculture and silviculturists are facing all over the world. As often is the case, the decision to write this book originated from discussions around a beer.

We thank Sybille Haeussler for reducing platitudes, Erin Hall for help with tables and figures, and Maureen Puettmann for help with references. Mike Papaik, Louise de Montigny, Roberick Negrave, Sierra Curtis-McLane, Tim Works, Daniel Gagnon, Rasmus Astrup, Susan Hummel, Juergen Bauhus, and several anonymous reviewers provided valuable comments on various chapters. Financial support from the Sustainable Forest Management Network of Canada for travel of the authors to meet is gratefully acknowledged. Numerous colleagues in forest research and teaching institutions across the world are thanked for generously sharing their perspectives and insights. Last, but not least, we acknowledge many productive discussions with graduate students.

Introduction

Our incentive in writing this book is driven by the dramatic change in public attitude toward forests since the 1980s (e.g., Langston 1995) and the increased understanding of the ecological importance of maintaining structurally and functionally diverse forests. As a result, forestry is undergoing a major transformation. However, the silvicultural systems, practices, and approaches currently applied by silviculturists are still based on the same philosophies that led to the development of silviculture in central Europe more than a century ago. Silviculturists are struggling to modify their practices to meet the changing public perceptions and demands (O'Hara et al. 1994; Messier and Kneeshaw 1999; O'Hara 2001; Burton et al. 2003; Gamborg and Larsen 2003). Weetman (1996, 3) puts it succinctly when he points out that "European silvicultural systems . . . did not evolve to handle . . . complexity" demanded of forest management in the late twentieth century and refers to nineteenth-century European silvicultural approaches as "ideas that . . . tend to linger beyond their time."

The entire philosophical approach to silviculture, including how silviculturists choose and apply individual practices, needs to be critically assessed during such times of change. It is especially important to examine how silvicultural practices are linked to a varied set of factors, such as economic interests, scientific understandings, and political trends (Büergi and Schuler 2003). It is healthy to question the suitability of current silvicultural concepts, assumptions, and practices in light of changing societal views of forests, our broader ecological understanding of forested ecosystems, and the potential impacts of global warming on forests.

A Critique of Silviculture: Managing for Complexity is aimed at complementing current books in the fields of silviculture and forest ecology. This book provides advanced students, professionals, ecologists, environmentalists, and the interested public with an understanding of the history of silviculture and why silviculturists have managed forests in a certain way, an overview of important ecological concepts, an appreciation of differences and similarities between silviculture and ecology, and a road map to a new philosophical and practical approach to silviculture that endorses managing forests as complex adaptive systems. We believe forestry in general and silviculture specifically will benefit greatly by adopting some of the key characteristics of the science of complexity. Forests are perfect examples of complex adaptive systems, and complexity theory suggests that integrating "complexity" into silvicultural prescription will enhance the resilience and adaptability of managed forests. This is of special relevance in the context of future climate change, as forests will likely be exposed to a new and different set of disturbances.

We focus our discussion on within-stand relationships since it is the scale at which many processes operate that silviculturists manage and it is where our expertise lies. Incorporating concepts of complexity science into silviculture will facilitate continuous production of the many goods and services society now expects from forests while improving on ecosystem resilience and adaptability in the face of climate change and other unexpected disturbance agents. In no way should this book be viewed as downplaying the crucial role of commodity production as a worthwhile management goal. As long as humans use wood and other forest products, production of these products will be a necessity. In fact, as we learn more about the environmental impacts (e.g., energy requirements, pollution, carbon balance) of the production and utilization of alternative materials, the use of wood may become even more popular.

Chapter 1 provides a historical perspective on the development of silviculture. It suggests that silvicultural concepts and practices are intrinsically linked to the specific economic, ecological, and political circumstances that led to their development and wide acceptance. Chapter 1 concludes that silvicultural approaches and practices can be properly understood only in their historical contexts.

Chapter 2 presents a critical review of the core principles that have formed the foundation of silvicultural thinking, study, and practice. The chapter examines how silviculture has focused on commercial tree species with an agricultural approach to research and practice, leading to silvicultural practices being applied uniformly at the stand-scale. Chapter 2 further explores how the desire for predictability has affected silvicultural practice and research and how it has encouraged a strong, top-down command-and-control approach to the management of forested ecosystems.

Chapter 3 reviews general concepts and theories in ecology with an emphasis on how the desire to understand ecosystem complexity has affected the development of the discipline. The chapter illustrates how the notion of complexity has always been implicit in the science of ecology and how this notion has influenced theories and tools used by ecologists to understand and study the natural world.

Chapter 4 contrasts the fundamental views and approaches of the disciplines of silviculture and ecology. These differences exhibit themselves in textbooks and the structure of research organizations, as well as in limited cooperation among their leading research organizations. We then discuss the movement toward large-scale management experiments in silviculture. We specifically focus on the inherent conflict between the core attributes of silviculture discussed in chapter 2 and the broader objectives of contemporary large-scale silvicultural studies to find ways to incorporate greater variability (structural and ecological) into silvicultural practice.

Chapter 5 contains our road map on how silviculture needs to change in order to manage forests as complex adaptive systems. We explain the origins of the science of complexity. This is followed by our "operational" understanding of forests as complex adaptive systems and the main challenges silviculturists face when managing for complexity. A comparison of the impacts of the even- and uneven-aged traditional silvicultural systems with that of a natural forest highlights how silvicultural practices can reduce the range of possible options that natural forests exhibit. We then cover the main subject of the chapter by reviewing how the core attributes of complex adaptive systems should be considered by silviculturists. Finally, we provide a list of steps that silviculturists can implement to move silviculture toward managing forests as complex adaptive ecosystems. If we are successful at convincing the reader to follow us down the "complexity" road, we expect that silviculture will be more effective at solving the breadth of future management problems, regain its *lettres de noblesse*, and also be more fun and fulfilling.

1

Historical Context of Silviculture

Scientific exploration and natural resource management occur in direct response to human need. Forest science and management are no exception. In this chapter, we review the history of human interaction with forests. In examining how social, economic, and ecological circumstances influence *silviculture*, we offer numerous examples in support of Cotta's observation: "There would be . . . no forest science without deficiency in wood supplies. This science is only a child of necessity or need" (Cotta 1816, 27). We show how the development and application of silvicultural concepts and practices involving the manipulation of forest vegetation to accomplish a specified set of objectives has been closely tied to natural resource issues pertinent to specific localities at specific points in time. Our focus is central Europe, where silviculture first developed (du Monceau 1766; Hartig 1791), and North America, which has adopted many European practices (Hawley 1921), because we are most familiar with these regions and their silvicultural literature. Despite the historic, cultural, and linguistic differences that influence specific silvicultural practices, our main arguments also apply to other regions.

Management approaches and silvicultural practices must be viewed within the context of contemporaneous economic, societal, and cultural developments (Weetman 1996). The general history of human relationships with forests has been extensively reviewed (Smith 1972; Mustian 1976; Thirgood 1981; Hausrath 1982; Mantel 1990; Kimmins 1992; Schama 1995; Weetman 1996; Botkin 2002). The variety of silvicultural practices is attributed to practices developing independently in multiple regions (Mayr 1984; Mantel 1990), indicating that small-scale, local conditions are important in understanding the historical context of silviculture. Just like any scientific development, the rate of change in silviculture has been neither linear, constant, nor even continuous (Kuhn 1962; Hausrath 1982; Mantel 1990; Bengtsson et al. 2000; Tomsons 2001). Instead, the progress of silviculture directly followed trends in societal developments. During periods of fairly constant social and environmental conditions, such as during the 1950s through the 1970s, forest management changed little. On the other hand, times of societal upheaval or transformation quickly resulted in fairly drastic changes in forest practices. Our definition of "societal development" includes changes in basic demands for commodities from the forest, improvements in scientific understanding of forest ecosystems, and changes in philosophical, cultural, and spiritual attitudes toward forests.

This chapter provides an overview of the history of forest management and silviculture because it is important to understand how silviculturists arrived at their current set of practices. Possibly even more important is the need to understand how the historical development of silviculture has affected the cultural attitudes of silviculturists and the way they think and address problems. It is the combination of historical convention and current scientific understanding that provides the basis for choices that so profoundly affect the management of forests. A basic understanding of silvicultural history provides useful and necessary context to the contemporary debate about the future role of silviculture in managing forests. We present a brief history of the external factors that were most influential on forestry and describe how human needs and external conditions led to the development of silvicultural practices and the subsequent combining of individual practices into *silvicultural systems* to meet management objectives. We highlight the importance of context, especially the need to consider time and place when evaluating practices, and discuss issues associated with "adoption without adaptation" by presenting examples of where silvicultural practices successful in one region were transplanted to other conditions or regions.

Major External Factors Influencing Development of Forestry and Silviculture

External factors are factors outside forestry that had a large influence on the field of forestry and the discipline of silviculture and originated from a variety of economic and social conditions. The main factors discussed in this chapter include population pressures, shifts in economic philosophy, development of industries, and scientific and technical advancements. The most important factor driving changes in forest management in central Europe during the last 2,000 years is the ever-increasing pressure of human populations on the natural resources. This pressure is determined through a combination of human population levels (fig. 1.1) and changes in the standard of living with an associated increase in the demand for forest products. For a brief perspective, during Roman times, the human population in central Europe was estimated to be less than 34 million. Settlements were separated by large tracts of forest, although they were not necessarily culturally or economically isolated (Schama 1995). Major trade routes existed, but larger population movements were quite limited, resulting in fairly stable population levels (McEvedy and Jones 1978).

For the last 2,000 years, the human population has increased at an ever-faster rate, with notable exceptions. Several famines (e.g., Great Famine of 1315–1317), disease pandemics (e.g., typhoid in 1309–1317, bubonic plague in 1348), and periods of intense warfare (e.g., Thirty Years' War of 1618–1648) not only slowed rates of population growth in Europe, but also were responsible for major population declines in many regions. Other societal developments, such as the emergence of new farming techniques, the appearance of potatoes as a human and animal food source, and improved medical knowledge, increased the rate of population growth. Emigration, especially the emigration wave to the Americas during the nineteenth century, slowed population growth in

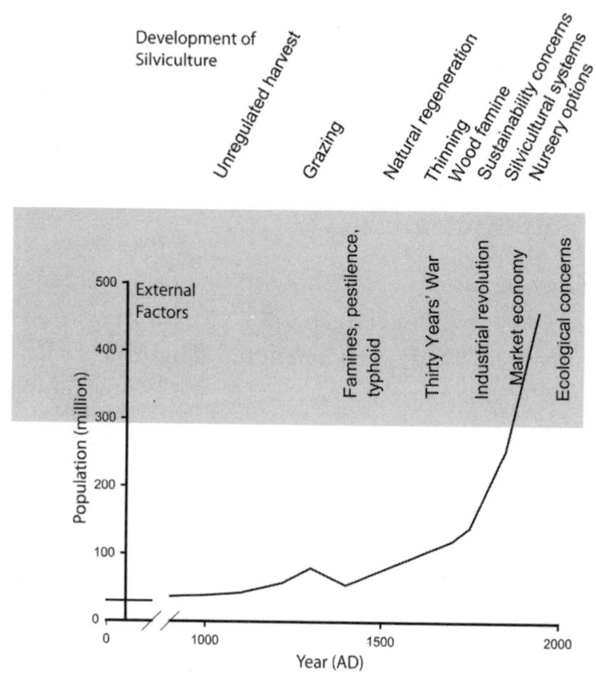

Figure 1.1. Historical population trends in central Europe (based on McEvedy and Jones 1978). Selected external factors that influenced the development of forestry are presented in the shaded area. Major factors that affected the development of silviculture are found above the shaded area.

Europe. More recently, the population in central Europe is decreasing (mainly due to low birth rates) but the impact of the declining population on the forest resource may be offset by an increased standard of living.

Major shifts in the economy of Europe in the late seventeenth and early eighteenth century strongly influenced the philosophical and cultural factors in the development of silviculture (fig. 1.1). During that time, economies in many parts of central Europe shifted from an agricultural base to an industrial base. The development of iron, salt, and glass industries in the sixteenth century caused a rapid increase in the demand for wood (Mantel 1990). The demand for energy wood, however, decreased somewhat in the eighteenth century as coal and oil replaced wood as an energy source in many factories. Other uses, such as the use of wood to support mine shafts and in the building of large shipping fleets, took its place.

Other influences of industrialization had a longer-lasting impact on the human relationship to forests and forest uses; some of these influences continue today. For example, hand in hand with industrialization

---

Economic Liberalism: Mainly attributed to Adam Smith (1776); assumes that economic activities are based on private self-interest and government has no role in shaping an economy. Viewed as the beginning of free-market economic liberalism, including free trade, capitalism, and libertarianism.

---

came new ideas about economics from philosophers, such as Adam Smith (1723–1790). Especially the development and popularization of *economic liberalism* and a free-market economy was novel to the forestry sector. While wood products have been traded for a long time, the appearance of wood in a marketing context is first documented in the fifteenth century (Lorey 1888). However, until the seventeenth century, forest products were either used locally or sold in markets that were regulated strictly by local principalities (Mantel 1990). The shift in economic thinking in the eighteenth century and the adoption of free-market ideas and concepts of management efficiencies by silviculturists (see chap. 2) had a long-lasting impact, and still influence our understanding of forest management and the development and application of silviculture today.

Forestry was rather slow to adopt economic liberalism, compared to other industries. But when it did, the view of the role of economics in

---

Normalwald (or *Normal Forest*): A normal forest is an idealized forest composed of even-aged, fully stocked stands with a balanced age-class distribution. The number of stands is a function of rotation length, as one age class is harvested and regenerated each year. Under the assumptions of similar species mixtures, site qualities, and tree densities and qualities, the normal forest has constant increment and volume and provides for a continuous supply of wood.

---

the ownership of forests changed dramatically. The forest had previously been viewed primarily as a stable component of a regional economy and employment base. Management decisions were applied in this context (Ruppert 2004). With the adoption of economic liberalism in the nineteenth century came the notion that the purpose of forests was to maximize profit for landowners (Ruppert 2004). This was a substantial shift in thinking, and its influence on forestry research and management activities cannot be underestimated. To apply the notion of profit maximization in forestry required new concepts and decision-making tools (Mantel 1990). In response to this demand, silviculturists started to inventory forests and document their growth and utilization (Hundeshagen 1826). The most notable advances in this context were the development of the *normal forest* concept (*Normalwaldkonzept*; Hundeshagen 1826; Speidel 1984) and the *Faustmann formula* (Faustmann 1849), both of which are still central ideas in forestry today (Speidel 1984; Edwards and Kirby 1998; Brazee 2001; Davis et al. 2001; Salo and Tahvonen 2002).

Faustmann Formula: Intended as a method to calculate the value of forestland. However, its main historical use has been to assess economically optimal rotation ages. The land expectation value (LEV) is calculated as:

$$LEV = \frac{P(t)V(t) - C}{e^{rt} - 1} - C$$

where $P(t)$ is the stumpage price of trees at stand age t, $V(t)$ is the stand volume at age t, C is the regeneration cost, and r is the interest rate. Stumpage price, volume, and regeneration costs are held constant and the response of LEV over stand age is used for assessment of rotation age.

Under economic liberalism, all forest management activities were viewed as investments and therefore subject to economic evaluations. The calculation of interest rates for management activities and forest properties (Cotta 1817) was especially prevalent in the *Bodenreinertragslehre* (Speidel 1984). Under this popular economic school, maximizing interest rates was the dominant consideration in the decision-making process. Viewing forests through this fiscal lens profoundly changed the foundation for silvicultural decision making. Previously, silvicultural decision criteria were based on the structure of forests as defined by volume or area of harvestable, fully stocked stands. These criteria were now replaced by productivity criteria, for example, current and expected tree and stand growth as reflected in profits. In practice, under this economic philosophy rotation lengths were fairly short, mainly due to the impact of interest compounding. For the same reason, fast-growing species were usually favored in regeneration efforts and management activities were implemented only if they either were cheap or resulted in quicker recovery of investments due to faster growth of the managed trees

Bodenreinertrag: An economic philosophy based on the belief that economic interest is the sole purpose of forest management. Management practices (on private and public land) are market-driven with the goal of maximizing the internal rate of return. Optimal rotation ages can be calculated with the Faustmann formula. The impact of interest rates in determination of profits leads to short rotations.

As with any trend, these new economic approaches were not accepted by all silviculturists, and alternative views developed. Especially, some silviculturists questioned whether using the internal rate of return as a dominant driver of forest management decisions was appropriate for an industry with long-term investments, such as forestry. Other ideas, most prominently the *Waldreinertragslehre* (Speidel 1984), became recognized as viable alternatives (Ruppert 2004). The management goal under the *Waldreinertrag* focused on maximizing annual returns rather than the internal rate of return. Since returns were calculated as the difference between investments and revenues on an annual basis, interest rates were not considered when evaluating the profitability of management activities.

Compared to the *Bodenreinertragslehre*, the *Waldreinertragslehre* encouraged implementation of more intensive forest management practices

Waldreinertrag: An economic philosophy that acknowledges the social responsibility of landowners to the greater community. Management goals include the maximization of annual profits. Since these are calculated without the influence of interest rates, optimal rotation ages are generally longer than under the Bodenreinertragslehre.

with little concern for the delay between when investment costs were incurred and recovered. One of the best examples of this philosophy is the management of high-value oak (*Quercus robur* or *Q. pubescens*) stands in central Europe, specifically in Spessart, Germany. Because of the extremely high value of quality oak logs, just about any investment can be justified under the *Waldreinertragslehre*. Typical practices in these stands include very expensive reforestation activities, such as dense planting, intensive vegetation control, and multiple pre-commercial thinnings, underplanting of European beech (*Fagus sylvatica*) or other trainer species, or artificial pruning (Burschel and Huss 1997). Moreover, without the compound interest penalty, longer rotations and associated management goals such as large, high-quality timber became more common. Typical rotations for oak in these regions vary between 150 and 240 years, a length that could never be justified under the *Bodenreinertragslehre* economic philosophy.

These two economic approaches became a widespread basis for forest management decisions, partially reflecting the different values that societies place on private property and social responsibilities. In Europe most emphasis was on the *Waldreinertragslehre*, while North American forest economists tended to favor the *Bodenreinertragslehre* (Speidel 1984; Davis et al. 2001). Over the years the two approaches were refined and modified, but their basic fundamental principles are still the dominant basis for forest management decisions on many ownerships today (Davis et al. 2001).

The influences of economic liberalism were so entrenched in the forestry profession and were so widely accepted that they carried across ownerships with different management objectives. In many regions, ownership patterns were not easy to detect just by examining forest conditions in the landscape (Ohmann et al. 2007; Spies et al. 2007). Public, small private, and industrial owners obviously had different management constraints and goals. These differences, however, were smothered by the common economically driven approach to forest management. The fairly homogenous landscape (in terms of stand sizes, rotation lengths, and harvesting patterns) partially reflects an educational system that did not directly distinguish between training silviculturists for different ownerships. Also in some regions, specifically in Germany, the line between public and private forestry was blurred; a typical job description of state forestry employees included not only management of state land, but also consultations with small, private woodlands. All of these aspects allowed a single dominant philosophical approach—that is, economic liberalism—to express itself by homogenizing the forested landscape.

This homogenization of forests of different ownerships did not change significantly until the 1990s, when the emphasis of management on public land shifted away from a focus on timber production. In many regions, especially in North America, public owners have moved from economically driven management approaches toward some form of ecosystem management with a focus on late-successional habitat and therefore longer rotations and partial harvests (Kohm and Franklin 1997). Industrial forestlands remain driven by economic incentives with fairly short rotations. Small private landowners appear to fall somewhere in between those two extremes, often focusing less on economic values and more on recreational and ecological values (Uliczka et al. 2004).

Another major factor that influenced the human relationship with forests was the progress in scientific understanding of forest ecosystems. During early human history, forest management efforts were limited to gathering wood products and tending the forest for agricultural use, such as animal grazing (Hasel 1985; Mantel 1990). However, during Roman times humans developed an understanding about regeneration requirements, specifically for sprouting and growth rates of different tree species (Hausrath 1982). During the next 1,800 years, much of the new scientific knowledge was locally developed and applied by foresters, whose main tasks were focused not on silvicultural applications, but on hunting and police functions. With few exceptions, most information was carried forward through oral tradition. In Europe, the first comprehensive documents demonstrating a scientific understanding of ecological and silvicultural issues were prepared by Hartig (1791) and Cotta (1817).

These publications can be viewed as the initiation of silviculture as a scientific undertaking. Shortly thereafter, the science of ecology became established (chap. 3) and investigations into ecosystem structure and function began, but they had little impact on silviculture for a long time (chap. 4). The establishment of research institutions in government agencies and forest faculties at universities (e.g., 1792 at Freiburg, Germany;

1805 at Koselev, Russia; 1811 at Tharandt, Germany; 1824 at Nancy, France; 1828 at Stockholm, Sweden; 1862 at Evo, Finland; 1870 at London, Great Britain; 1898 at Biltmore and Cornell, United States; and 1900 at Yale University, United States) is a clear sign that forestry, and thus silviculture, had become a recognized scientific discipline.

Parallel to the development of a scientific understanding of forests and forestry, technological advances greatly impacted the choice of silvicultural practices. Examples of technologies that directly impacted forest management include metal axes, crosscuts (early twentieth century), and chainsaws (around 1950), and starting in the 1970s harvesting machines such as feller-bunchers. All these tools, in conjunction with improved transportation technologies in the twentieth century, allowed more efficient cutting and therefore harvesting of wood.

The preceding discussion outlined the main external factors (population pressures, shifts in economic philosophy, and scientific and technical advancements) that, taken together and in conjunction with other factors (too numerous to describe here), defined what Cotta calls "human necessities and needs" and opportunities to fill these needs. Silvicultural approaches and individual silvicultural practices can be properly understood and evaluated only within this broader societal context. As external factors changed, demands and opportunities for forest management also changed, creating new management objectives and constraints, resulting in new silvicultural practices (fig. 1.2). An appreciation of how silviculture evolved under these pressures is crucial for understanding how silviculture is being conducted today and is also very helpful in discussions about the future of silviculture. Many consequences of silviculture's response to pressures of industrialization and population growth—such as the development of silvicultural systems, the refinement of nursery operations and planting practices, the predominance of conifer regeneration, and shorter economically driven rotations—remain visible in the landscape today.

The Development of Silviculture

The historical development of human societies, forests, and the management of forests are strongly intertwined (Diamond 1999; Farrell et al.

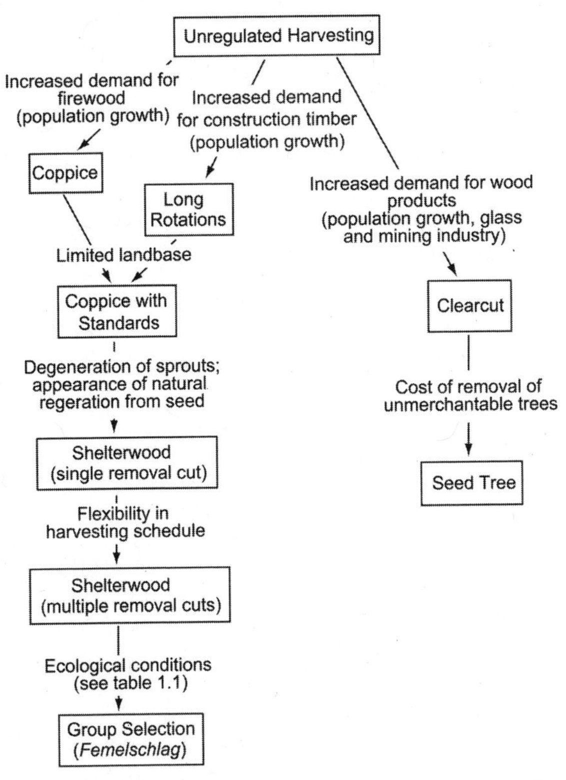

Figure 1.2. Simplified diagram highlighting major factors that influenced the development of silvicultural systems in central Europe. Note that this is not a timeline representing the use of the systems.

2000). Humans have actively manipulated their environment for millennia to fill their "needs and necessities." In Europe, the earliest documented human impact on the landscape is extensive land clearing for agriculture dating back approximately 5,000 years (Schama 1995; Burschel and Huss 1997). At that time, forest management was limited to utilization such as firewood gathering. As the human population increased, keeping livestock became more common and forests became a place for grazing or herding (Hausrath 1982; Mantel 1990). The technology to harvest, transport, and utilize wood was not well developed, and wood harvesting was only of minor, local interest. Exceptions to this included areas near waterways, where early societies could utilize forests extensively to support a large shipping fleet. Consequently, early management activities consisted mainly of burning forest parcels selected for clearing. The goal of this practice was to open forested areas sufficiently to support animal grazing or field crops (Hausrath 1982; Hasel 1985; Mantel 1990). Apparently, humans already understood that open forest conditions led to vigorous herb and shrub layers and that management was necessary to maintain such conditions (Bengtsson et al. 2000). The practice of burning to enhance forage also encouraged seed production from shrubs and trees, a major source of food for animals and humans.

Although limited in extent, these early management practices were reflected in the appearance of the forested landscape (Bengtsson et al. 2000). For example, in the central European hardwood region, burning and clearing led to open forest conditions with scattered large crowned oaks and beeches. Other hardwood and shrub species were often relegated to the understory. Because of the effort involved and the limited infrastructure and transportation technologies, such practices were not applied homogenously across the landscape, but were concentrated around population centers. During this period, forests near human settlements were quite heavily impacted, while forests farther away from human settlements remained essentially unmanaged (Hausrath 1982).

At the end of the first millennium AD, the utilization of forests expanded from grazing animals to a greater emphasis on wood production as the European population expanded quickly and agriculture became more efficient at providing food (Mantel 1990). Previous management practices, such as burning to clear forestland for grazing, could not accommodate this shift in emphasis. Newer, more intensive management practices had to be and were developed to provide a consistent supply of wood products. A greater diversity of management practices emerged. For example, to supply firewood and construction timber simultaneously, silvicultural practices included repeated cutting of small hardwoods to produce firewood through coppicing and cutting the largest and best trees for construction timber (Hausrath 1982; Hasel 1985; Mantel 1990). Despite these efforts, it soon became apparent that limiting forest management to zones near population centers would not meet the higher demand in many regions.

The deliberate efforts to expand forest management outside the direct vicinity of settlements could be interpreted as the beginnings of landscape management (Hausrath 1982). During this early medieval period, choices of management practices at the landscape level were a result of the diversification of objectives that included an increased demand for grazing areas, pressures to provide hunting opportunities and thus habitat for game animals, and production of a diverse set of forest products. This is quite different from earlier landscape decisions, which were driven by the challenge of managing for multiple products simultaneously. During Roman times, for example, landowners already distinguished between areas managed for firewood (*silvae caeduae*, or coppice forests) and those managed for food production (*silvae glandairae*, or forests dominated by trees with nutritious seeds that could be used as animal feed) (Hausrath 1982).

Another shift toward a stronger focus on wood production was driven by industrialization in central Europe. Forestland use in the Middle Ages included efforts to facilitate grazing of farm animals or provide habitat to support game hunting. The development of industries in the nineteenth century that relied on wood went hand in hand with the appearance of more efficient agricultural techniques and crops and the loss of hunting privileges by royalty. In some regions, expansion of a specific market for wood—such as firewood for the iron industry in southern Northrhine-Westfalia, Germany—was not compatible with the production of other forest products. Here, the zoning approach reflected itself in intensive management of stands for single products. In most parts of Europe, the variety of desired wood products, such as small and large construction timber, was more compatible and forests could be managed for multiple products. Therefore, during the later Middle Ages, a combination of thinning and final harvesting operations, often in conjunction with mixed-species management, ensured the supply of a diverse range of products (Hausrath 1982; Hasel 1985; Mantel 1990). The high alpine regions in central Europe are a noteworthy example of an area with a dominant non-wood-related objective. Forests in these regions were specifically managed to provide continuous avalanche protections for settlements, while providing timber and firewood was secondary

(Schönenberger 2001). Since avalanche protection was achievable only with continuous forest cover, these areas became a major force in the development of uneven-aged silvicultural systems.

Fundamental Concepts and Practices that Influenced Silviculture

Inventory and Planning

The increased human population and industrial demand for wood resulted in the application of industrial thinking in forest management, including the use of inventory and sophisticated planning procedures. The late seventeenth century was the beginning of a period of rapid change for the inhabitants of central Europe, and forest management went through a period of intense transition to accommodate these changes. After the Thirty Years' War (1618–1648), the increased human population and demands of the emerging mining, glass, and ship-building industries had led to desolate forest conditions in many regions. Forests with low timber volume and value, and areas degraded to the point where they no longer supported trees, became a common sight in the central European landscape (fig. 1.1).

These conditions led to the first published discussion of wood supply sustainability (von Carlowitz 1713). They also led to the development and use of inventory and forest planning tools (Hartig 1795; Cotta 1817), which became widely adopted. The tools were so successful that inventory and planning (forest regulation) became a dominant field in

Fachwerkverfahren: A forest management approach in which forests are divided into similar-sized management units. The goal of this division is to ensure a long-term supply of wood and stable age-class distributions (see *Normalwald*). Units were selected to have equal area (*Flächenfachwerk*) or harvesting volume (*Massenfachwerk*).

forestry sciences in the mid-nineteenth century (Mayr 1909; Morgenstern 2007) and have continued to be influential in the development and implementation of silvicultural practices (Mantel 1990). While concepts of cutting units and cycles had been used regionally for some time, the new inventory systems and their associated mathematical advancement allowed formal assessments and planning. For example, methodologies such as the *Fachwerkverfahren* enabled the calculation of "sustainable" harvest levels (Hasel 1985; Mantel 1990; Morgenstern 2007).

More important for silviculture, these new planning tools became the criteria used to assign harvest operations to specific stands, replacing the silvicultural analyses of individual forest conditions. Thus, many silvicultural decisions such as layout, size, and timing of harvest operations were now driven by economic or planning priorities rather than by site-level ecological conditions. The increased interest in economic liberalism and focus on productivity helped to spread these ideas, and planning procedures have dominated forestry operations on many ownerships ever since (Speidel 1984; Davis et al. 2001). Already Mayr (1909) complains that the sizes of stands or inventory units are not based on ecological considerations. He suggests that ecological criteria would lead to management of "mini-stands" of 0.3 to 3 hectares, a size that would "fit most forest types in the world."

As the normal forest concept started to be reflected in forest planning and regulations procedures, it influenced silviculture to such an extent that its consequences are still reflected in today's silvicultural practices and are easily visible in the landscape today. The normal forest was developed as a conceptual model for organizing ideas about growing stock and growth and yield relationships (Hundeshagen 1826; Heyer 1841) and to calculate sustainable harvest levels. However, forest managers did not limit the use of the normal forest concept to its intended use, but instead interpreted a normal forest as a desirable goal for forest ownerships (Speidel 1984). This second interpretation of the normal forest concept ensured that the underlying assumptions of the concept had a major influence on how forests were perceived and managed. The assumptions included that forests were composed of units (i.e., stands), which were (1) homogenous in species mixtures or monocultures, (2) homogenous in size and site conditions, (3) fully stocked, or with homogenous stocking, (4) of homogenous wood quality, (5) organized spatially to facilitate harvesting, and (6) without risks of natural damage and catastrophes. Consequently, forest regulation and planning efforts aimed to transform the natural forest into a collection of homogeneous stands with a balanced age-class distribution (see also discussion of the Faustmann formula). The desire to achieve normal forest conditions became a dominant management goal, and silvicultural approaches and practices were evaluated based on how they helped achieve this goal. Over time, the normal forest concept has been expanded to include other aspects such as risks management (Klocek and Oesten 1991). Concerns about improper applications of this concept, such as using the normal forests as desired management goals without fully acknowledging the assumptions, are still prevalent today.

The impacts of the normal forest concept on the development of silvicultural systems and practices during the past 150 years cannot be underestimated. In this context, Weetman (1996, 12) suggests that "most European silviculture has been oriented towards sustained yield" and that "the principle of sustainability has inspired all silvicultural systems." The legacy of the normal forest concept was a strong focus of silvicultural approaches and practices on fully stocked stands, stands with fairly simple structure and composition, intensive thinning practices, and harvest timing determined by productivity measures.

Even small-scale harvesting patterns, such as the single-tree or group selection systems (described in detail later), are conceptually based on the normal forest concept (Mantel 1990). These systems were not developed to match the scale of management activities to the scale of ecological functions and processes. Instead, in their basic approach, uneven-aged selection systems are conceptually very similar to even-aged silvicultural systems. Both aim at ensuring a sustainable supply of wood by determining when an individual stand or tree needs to be harvested (Mantel 1990).

The selection systems developed as silvicultural systems from unregulated high-grading (*Plenterung*) in conjunction with development of an inventory or planning system, the so-called control or check method (*Kontrollmethode*) (Biolley 1920). Even though Ammon (1955) pointed out the limited applicability of the normal forest concept to single-tree selections, the development of the control method was influenced by the normal forest concept, just as even-aged silvicultural systems were (Mantel 1990). The control method, however, focused on obtaining the highest yields within individual stands, with less emphasis on achieving "normal" growing stock (Mantel 1990). Thus, by following the *Waldreinertragslehre* or *Bodenreinertragslehre*, silviculturists using the control method still based management decisions about stand structures, size distributions, growth patterns, and their impacts on the goal to maximize forest productivity. Claims that the initial development and use of single-tree selection as a silvicultural system were driven by the desire to maintain natural stand structures and dynamics and within-stand variability for ecological reasons cannot be substantiated (Mantel 1990).

Species Mixtures and Monocultures

Discussions about the benefits of mixed species and monoculture management have been influenced by more than an understanding of ecological issues. External social factors and management constraints have had just as much impact, or maybe even more, on silvicultural decisions about species makeup of managed forests (Hausrath 1982; Hasel 1985; Mantel 1990). Natural forests in central Europe commonly contained multiple tree species, and early management efforts did not pay particular attention to or even affect species composition. The first notable exceptions were large, wide-crowned oak and beech that were favored to ensure high seed production as nutritious food for animals (Mantel 1990). In the sixteenth and seventeenth centuries, as management practices intensified, the choice of species and the question of monocultures versus species mixtures became topics of intense discussion (Hausrath 1982). Early writings suggest that maintaining or duplicating species mixtures

Kontrollmethode or Control Method: Developed by Gurnaud in the late nineteenth century and refined by Biolley in the early twentieth century. This planning method is based on continuous inventory of tree growth patterns. Inventories stratify the growing stock by size classes. Management decisions are based on comparison of current with "ideal" size class distributions.

found naturally in the forests was considered best for providing a sustainable wood supply (von Carlowitz 1713). Despite best efforts, however, management practices (e.g., natural regeneration, thinning) were not refined enough to achieve and maintain desired species mixtures throughout the life of a stand (Mantel 1990). By the end of the eighteenth centrury, Hartig (1791) voiced concerns that differential growth rates and competitive abilities would lead to forests that were dominated by a single species. The lack of success of mixed-species management was subsequently used as a justification to manage for single-tree species (Hausrath 1982; Mantel 1990).

The onset of economic liberalism and its focus on productivity further strengthened the trend. The shift toward monocultures had a variety of impacts on silvicultural approaches and practices. For example, it required better control of regeneration than when managing for mixed species and was therefore at least partially responsible for development of better artificial regeneration methods. Also, large-scale shelterwood and clearcuts became more common starting in the 1820s, in part because they were more suitable for establishing monoculture stands (Hausrath 1982). Over time, management of monocultures became a standard practice in central and northern Europe, but critics started to voice concerns about this trend in the late nineteenth century (Gayer 1886). As scientific understanding of regeneration methods and growth patterns increased, interest in management of mixed-species stands revived (Cannell et al. 1992; Kelty et al. 1992). However, for a long time, discussion about the benefits of single versus mixed-species management focused almost entirely on growth and production (Assmann 1961). The ecological benefits and values of multispecies stands have only recently become of interest (e.g., Berger and Puettmann 2000).

Stand and Rotation

The stand concept is a key feature that has allowed silviculture to be successful in the past. *Stands* are defined as a homogenous vegetation unit or "group of trees . . . that foresters can effectively manage as a unit" (Nyland 2002, 2). Starting with the first human harvesting activities, logistical constraints (tree sizes and infrastructure) in conjunction with complex and diverse forest conditions commonly resulted in the cutting of dispersed trees (Hausrath 1982; Hasel 1985; Mantel 1990). If even-aged cohorts were present, harvesting was concentrated in small groups. Because of the great effort required for cutting the forest, harvesting was usually done as a direct response to a need for a specific wood product.

Consequently, harvesting of larger units was inefficient in forests that were diverse in tree species, size, and quality. Many of the trees cut in larger units (what we now call stands) would not have been utilizable (see also discussion of clearcutting).

Harvesting activities became more concentrated in the Middle Ages, as tree regeneration became an important consideration for foresters when determining harvesting layout (Hausrath 1982; Hasel 1985; Mantel 1990). Specifically, the shift toward stands and management of stands was initiated (1) because of the inability to regenerate new trees under high grazing pressure by wildlife and farm animals, (2) to increase harvesting efficiency, or (3) for inventory and planning purposes (Hausrath 1982), and not because stands were logical, ecologically defined management units. Instead, protection of regenerating trees through hedges, fencing, and regulatory restrictions of farm animal grazing were the only feasible options to protect regeneration at the time. Spatially concentrating the area on which the regeneration needed to be protected—that is, the harvested area—made fencing and other protection efforts feasible and/or cheaper. Over time, the advantages of the stand concept, beyond the simple necessity of protecting the regeneration, were recognized (Mantel 1990). Because of higher efficiency of mapping and inventory, infrastructure, and concentration of planning and management activities, dividing forests into stands became widespread and now a globally established concept in silviculture (e.g., Kellomäki 1998; Fujimori 2001; Nyland 2002).

More than any other concept, the stand concept has been widely accepted as a basis for silvicultural decision making. It even encouraged the development of a subdiscipline, stand development (Oliver and Larson 1996), which expanded the stand concept to include dynamic aspects. The notion of "cutting cycle or rotation" is a second example of a long-standing, prominent silvicultural concept that has undergone changes throughout history in response to a variety of external factors. During the first millennium, rotations were determined by the desire for a certain forest product, which in turn was a function of product use, cutting tools, and transportation options. Thus, early cutting cycles for firewood in central Europe's hardwood region were rather short, for example, three to seven years (Hausrath 1982). Later, typical firewood rotations were lengthened to up to twenty years, and up to thirty or forty years for construction timber. Alternatively, in regions with a leather tanning industry, rotations were determined by the conditions of oak bark, rather than tree size (Mantel 1990).

With the onset of economic liberalism in forestry during the early nineteenth century, the determination of rotation ages shifted from a "product driven" to a "productivity driven" basis. Ideas about growth, growing stock, and sustainable yield (see discussion of the normal forest concept) expressed themselves in calculations of rotation ages and were commonly applied by landowners in Europe and North America (Speidel 1984; Mantel 1990; Davis et al. 2001). Only since the 1990s have silviculturists, especially on public land, reassessed the basis for calculation of rotation ages.

Regeneration

Tree regeneration has always been viewed as the most important task for silviculturists and an essential element of sustainable forest management (Lavender et al. 1990; Burschel and Huss 1997; Smith et al. 1997). The view of tree regeneration by silviculturists has changed dramatically over the past 2,000 years in response to external factors. Throughout history, tree regeneration was of concern to silviculturists only during times and in regions with wood shortages (Mantel 1990). The development and application of regeneration practices was directly linked to specific ecological, economic, and social conditions. Because of the relative ease of regeneration through vegetative reproduction, paired with the demand for grazing opportunities and firewood, coppicing is one of the oldest forms of managing regeneration. Already applied during Roman times, coppicing was used extensively in central Europe starting in the fifth to seventh centuries and retained a dominant status for several centuries (Hausrath 1982).

The first efforts at developing artificial regeneration practices by direct seeding, or the planting of seedlings or cuttings, came as a response to the practice of selected tree species being grown in specific locations, such as near settlements for shelter or as markers of political or property borders. Artificial regeneration started to be considered a tool to regenerate larger areas such as recently harvested stands (Mantel 1990) only at the beginning of the fifteenth century. For the next few centuries (up to the middle of the eighteenth century), artificial regeneration was applied sporadically; in many regions, simply relying on sprouting or the occurrence of natural seedlings was the dominant form of regeneration "management" (Mantel 1990).

Economic liberalism (late eighteenth and early nineteenth centuries) had a profound effect on the expansion of artificial regeneration efforts. The unreliability of obtaining natural regeneration, often despite great efforts, was considered unacceptable. Even more important, the perception of a standard or acceptable speed of reestablishment (and associated economic returns) changed. The fast establishment and growth obtained by conifer plantations became the "standard" expectation of economists. At the same time, increased demand for food (consider the famine of 1816) resulted in the need to grow food on marginal agricultural land and forested sites. Widespread applications of agroforestry practices encouraged artificial regeneration, and it became common that forests were cut and farmed for a few years before being abandoned again. Farming practices such as plowing, grazing, seeding, and harvesting of food crops eliminated or damaged natural regeneration; consequently, artificial regeneration was seen as the only viable option for reestablishing forests on these sites, and therefore the practice expanded (Hausrath 1982; Hasel 1985; Mantel 1990). In conjunction with clearcut harvesting operations, artificial regeneration became widespread in central and northern Europe and North America.

Thinning

Thinning aims to reduce stand density with the goal of improving the growth of residual trees, enhancing forest health, recovering potential mortality, or obtaining income. Thinning provides an example of a practice that has been implemented and modified over time in response to a variety of economic and ecological issues. Beginning in the fourteenth century, the need to harvest trees of various sizes began to be reflected in the interest in thinning activities. Traditional harvest patterns of cutting single or groups of trees were unsuited to fulfill this demand, especially after the widespread establishment of even-aged stands. Compared to earlier forest conditions, which provided a variety of tree sizes and species, the newly established even-aged stands were more homogenous. Without intermediate entries, the long period between consecutive harvests created wood shortages (Haurath 1982; Hasel 1985; Mantel 1990). Small trees, in particular, were in great demand for fence construction, firewood, or to support grapevines. It was the demands for specific wood products, rather than the ideas about increasing growth and vigor of residual trees, that led silviculturists to implement thinning practices in young, dense stands.

By the sixteenth century, the effects of thinning on tree and stand growth were better understood and thinning became a common tool used by progressive silviculturists in central Europe. The devastating social and economic impacts of the Thirty Years' War, however, affected forestry practices in Europe. Thinning, like many other progressive ideas, was abandoned and practically disappeared during the next century. Despite the potential benefits for increased stand growth and vigor, thinning was even outlawed by many local regulations or tolerated only in times of the greatest wood famine (Haurath 1982; Hasel 1985; Mantel 1990). Finally, approximately a century and a half later, progressive and influential personalities such as Hartig (1791) reestablished thinning as an acceptable silvicultural practice. With the help of influential supporters and an increased understanding of the silvicultural and ecological effects of thinning on forest dynamics, thinning quickly became a common silvicultural practice throughout central Europe. In contrast to the earlier interest, the revival of thinning as a silvicultural practice was driven by a better understanding of the impact of thinning on the overall growing conditions for residual trees, rather than as a mechanism to fulfill a specific need for wood products (Mantel 1990).

Artificial pruning provides another example of a management practice that has been continued over time in response to changing external factors. In a similar development to thinning, interest in artificial pruning first developed as a response to increased need for wood and later became a tool for management of wood quality and regeneration. Just as we can observe today in many developing countries, these early pruning efforts were aimed at providing small-diameter wood for home fuel

(Mantel 1990). Even though wood supply ceased to be a concern in developed countries a long time ago, pruning is still a standard practice taught in silviculture classes (Burschel and Huss 1997; Smith et al. 1997; Nyland 2002). The modern justification is on improving wood quality with a side benefit of improving light conditions in the understory for tree regeneration. As such, in regions with intensive forest management, like central Europe, pruning is still perceived as a sign of good and modern forestry practices.

Development of Silvicultural Systems

There is probably no single subject better suited for assessing the impact of external factors on silviculture than the development of silvicultural systems (fig. 1.2; for a detailed description of the systems see Troup 1928; Matthews 1989; Helms 1998; and other silviculture textbooks). All silvi-

Silvicultural systems are a set of basic management practices to regulate stand structure and species mixtures. They are labeled after the reproduction cutting method, but include all aspects of stand management. Even-aged systems promote regeneration of closely aged trees. The *coppice system* regenerates the forest from sprouts or root suckers of cut trees. The regeneration develops in a fully exposed microclimate in the *clearcutting system* after removal of all trees from the previous stands. In the *seed-tree system*, cuts are similar to the clearcutting system, except that a small number of canopy trees are left to provide seed. Seed trees are removed after regeneration is established. In the *shelterwood system*, regeneration develops beneath the moderated microenvironment provided by residual shelter trees, typically because of frost or heat concerns. Shelter trees are removed when regeneration is sufficiently large to withstand microclimatic conditions. The *group-selection system* is a method of regenerating uneven-aged stands in which trees are removed, and new age classes are established, in small groups. The *single-tree selection system* is similar to group selection, except that individual trees of all size classes are removed more or less uniformly throughout the stand.

cultural systems began as a set of practices in response to localized, site-specific needs and ecological conditions. Once a set of practices proved successful in fulfilling local needs they were often applied regionally (e.g., Bavarian *Femelschlag*) or even globally (e.g., clearcut and shelterwood systems). It is important to differentiate between the following

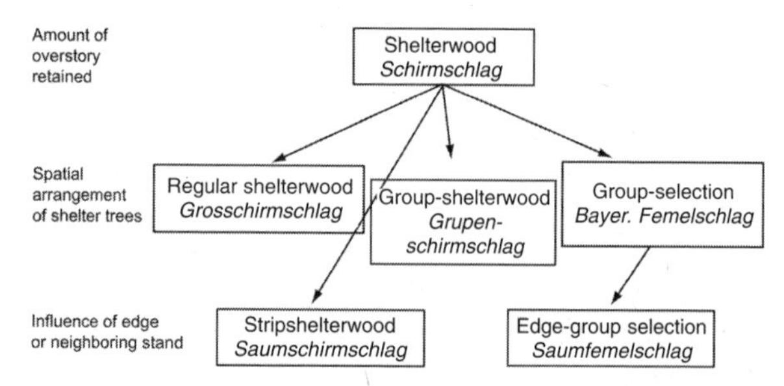

Figure 1.3. Example of how silvicultural systems were classified and documented by early German silviculturists. The shelterwood system was composed of a set of descriptive hierarchies. Descriptive criteria are given on the far left. (Adapted from Mayer 1984.)

two steps: (1) the development of a locally adapted set of practices, and (2) the expansion of these practices as they metamorphosed into a system for establishing, tending, and harvesting forests. The interplay between these two components was influenced by a variety of factors, not the least of them being the personalities involved.

Until the eighteenth century, silviculturists relied on experience (mostly verbal) and on the analysis of local social, economic, and ecological constraints and conditions (Hausrath 1982) to select their silvicultural practices. They simply did not have a commonly accepted, documented "toolbox" to assist in the selection. When constraints or conditions changed, silviculturists had no choice but to modify their local practices, usually by trial and error, to fit the new constraints or conditions. This led to the development of a wide variety of localized practices (Mantel 1990). By the late eighteenth century, as forestry and silviculture developed into an established discipline, silvicultural practices, including silvicultural systems, began to be classified and documented (Mayr 1909) (figs. 1.2 and 1.3). During the same period, universities and other schools first offered opportunities for a formal education in forestry. The formal education ensured that foresters were aware of the full variety of common silvicultural practices, but it also meant that for educational purposes these practices had to be categorized. As part of their education, foresters were taught new, modern technologies and practices and then trained to select from this set of management practices.

Despite educational needs to categorize, the variety of local ecological, economic, and social conditions in Europe resulted in the widespread application of a few dominant silvicultural systems in the nineteenth century (Spurr 1956), but a large number of modifications of these systems (mostly small-scale and locally applied) were still recognized. For example, Mayr (1909) lists fifty spatial and temporal modifications of silvicultural systems. A major challenge for the silviculture discipline has since been the development of an overarching set of principles and strategies that could encompass the diversity of practices without sacrificing the heterogeneity that arose from local ecological, economical, and social conditions.

The issue was resolved in central Europe through a classification system for silvicultural systems that included a hierarchy of criteria (Mayr 1909; Dengler 1930; Mayer 1984). At the highest level, the main descriptive criteria were the amount and timing of overstory removal (see fig. 1.3 for shelterwood example). Silvicultural systems were further divided based on spatial arrangement of residual trees, specifically whether a method was applied evenly throughout a stand or in large or small groups (e.g., group shelterwood or *Gruppenschirmschlag*). Another level was based on the influence of neighboring stand conditions (edge shelterwood or *Saumschirmschlag*).

The extensive list of possible combinations at these three levels allowed all localized systems to fit within the hierarchy. The classification system, rigorous but at the same time open, found general acceptance as one of the key concepts central to the discipline of silviculture (Mayer 1984; Burschel and Huss 1997; Fujimori 2001; Nyland 2002).

With a focus on local conditions in the nineteenth century, developing an inherently consistent naming system for the diversity of silvicultural systems that could be applied to different regions provided a challenge. It was solved by the development of a labeling system that made direct reference to the locality and specific conditions where the system was first developed and implemented (for examples, see Mayr 1909).

A prime example of "education leading to standardization" is the early silvicultural experience in Canada and the United States (Weetman 1996; Graham and Jain 2004). Many early North American foresters such as Bernard Fernow (1851–1923) and Gifford Pinchot (1865–1946) were trained or heavily influenced by European silviculturists. Consequently, when the North American forests began to be actively managed, the first silviculturists naturally turned toward the European silvicultural systems as potential options for managing their forests (Hawley 1921; Weetman 1996; Graham and Jain 2004). A review of early North American silviculture textbooks (e.g., Hawley 1921) reveals that they were very similar in structure and content to European textbooks. In fact, most of the photos in Hawley's textbook show the forests of central Europe.

The early descriptions of silvicultural systems in the North American literature attempted to cover the diversity of silvicultural systems, especially the variety of spatial modifications such as uniform, strip, group, or single-tree scales (Hawley 1921). However, in the translation, these silvicultural systems lost their ecological and historical context (Spurr 1956; Weetman 1996). This was especially critical since many readers in North America were not familiar with the conditions in central Europe that led to the development of these systems in the first place. In this transition, silvicultural systems were simplified to abstract practices, and the crucial role of adaptations of these systems to local ecological, economic, and social conditions was lost. For example, the main distinctions between local conditions were sometimes expressed in a prefix with regional names, such as Baden or Bavarian for *Femelschlag*. Obviously, the information in the label, and thus the need for labels, was lost to North American foresters, who were unaware of the particular and local conditions in these regions.

In essence, early foresters in North America were taught that they did not need to start from scratch and did not have to go through the process of assessing ecological, economic, and social conditions to develop their own locally adapted silvicultural systems. European systems were considered viable options for North America, and a major task of educated silviculturists became to select which one of these systems to apply in the various forests of the new world. In many cases that decision was even further simplified by the fact that the complex ecological, economic, and social relationships that were involved in the development of silvicultural systems in Europe were not present in North America.

Additional elements, specifically differences in language and associated difference in perception and understanding, played into the trend to simplify the choice of silvicultural systems in North America. The comparison of the understanding of the diversity of silvicultural systems in central Europe and North America provides an example of issues related to communication and perception. Spatial and temporal subtleties are accommodated in the German language by name-compounding (i.e., combining two or more words into a single compound word). When Spurr (1956) acknowledged the similarities between German and American silviculture, he highlighted this linguistic phenomenon as a major exception.

Spurr (1956) claimed that the American literature oversimplified silviculture practices and suggested that differences between the German and English language, especially the use of compound words, were a major contributing factor to this simplification seen in North America. A good example is the *Keilschirmschlag*. This single German word describes a silvicultural system that includes the utilization of an edge effect, with the edge shaped into a wedge (*Keil*). The second component of the word is a shelterwood (*Schirmschlag*). In terms of application, the implementations of wedge and shelterwood cuttings are not applied simultaneously, but follow a time sequence determined by regeneration success. Translation of this single word into the English language requires a lengthy and detailed explanation. Thus, descriptions of complex silvicultural systems were made more difficult in the English language. Cumbersome wording and associated difficulties in communication and perception were at least partially responsible for the loss of subtle distinctions in silvicultural systems (Nyland 2002). Maybe even more important, the understanding that the main silvicultural systems need to be assessed in the local or regional context was partially lost in the translation of the silvicultural systems from the German to the English language.

The power of nomenclature in driving the development of silvicultural practices can be seen by the experience in the Pacific Northwest of the United States. In an effort to encourage a rapid transition to sustained-yield management in the 1930s, Kirkland and Brandstrom (1936) suggested implementation of selective cutting. Selective cutting is a term that had (and still has) no specific definition, but has been applied to any kind of partial harvesting. It is quite distinct from the silvicultural systems labeled selection cutting (group or individual tree selection) (Curtis 1998). Despite Kirkland and Brandstrom's emphasis on a flexible application of selective cutting, including small clearcuts for the successful regeneration of Douglas fir (*Pseudotsuga menziesii*) and subsequent thinning in younger stands, their work was quickly interpreted as an unsuccessful attempt to implement the selection system in old-growth forests (Foster 1952; Isaac et al. 1952; Smith 1970, 1972). This misinterpretation, at least partially due to the practice of viewing silvicultural systems as simple categories, has been used to promote clearcutting as the only suitable silvicultural system in these West Coast Douglas fir forests (Doig 1976). This is partially responsible for the abandonment of silvicultural systems other than clearcutting in the northwestern United States. Furthermore, it has stifled research into other silvicultural systems for decades and is still influencing the discussion about feasibility of alternative silviculture practices in these forests (Curtis 1998).

Coppice and Coppice with Standards

The coppice method of regeneration has been a well-established management method in the hardwood forests of Europe since Roman times. The coppice method has been applied over time in a variety of ecological conditions with few changes because of the continuing need for small wood. Pressures to provide multiple products within a limited land base built up in the fifteenth century. Individual trees within a dominantly coppiced stand were preserved from harvesting so their seeds could be used as feed for pigs and their wood for large construction timber. This management approach was called "coppice with standards." It was very successful and became the dominant silvicultural system in many regions in central Europe. For several centuries, refinements of this system were limited to discussions of the optimal amount and spatial layout of standards (Hasel 1985; Mantel 1990). With a few regional exceptions, coppice with standards is no longer in use. The main reason for its demise was the drastic reduction in the need for firewood and the increased need for construction timber. Also, new silvicultural techniques such as thinning supplied markets that demanded smaller-sized timber, and mechanization required simplification of the spatial arrangement of the trees (Mantel 1990). The increasing use of artificial regeneration reduced the reliance on regeneration through sprouting. The coppice with standards method has been systematically replaced by even or uneven-aged forest.

Clearcut System

The development of the clearcut silvicultural system was closely linked to the onset of industrialization and economic liberalism in the seventeenth and eighteenth centuries (Mantel 1990). Industrial wood use meant that regional or national market forces, rather than local population needs, determined the demand for wood products. This shift, in conjunction with the onset of economic liberalism, required more systematic forest management practices, leading to intensified harvesting operations with larger harvest units. It was the beginning of large clearcutting operations. This pattern was not limited to central Europe. Over the last decades, numerous other regions in the world, including the Pacific Northwest of the United States, Canada, Russia, Asia, and the Amazon, have responded to the introduction of a regional demand for industrial wood by converting from opportunistic logging practices to large-scale clearcutting.

The shift toward a regional (and later national and international) marketplace for wood products about 200 years ago also implied that forested areas were not necessarily near locations where industrial wood demands were highest. Limited transportation options focused harvesting activities on mountainous terrain, such as the Black Forest and Alps, where fast-flowing rivers provided opportunities for log rafting. To justify the transportation expense, the size of harvesting units increased and utilization was maximized, leading to larger clearcutting operations (Hasel 1985; Mantel 1990). Early regulations called for the complete clearing of all trees from logged sites, including the removal of all small material to improve local grazing opportunities and avoid damage of residual trees to future regeneration (Hausrath 1982).

Seed-Tree System

Labor costs and logistics were primary drivers leading to the development of the seed-tree method. Not unexpectedly, loggers were reluctant to cut and move unsaleable material in the early industrial clearcut operations. These clearcuts became very difficult to implement, and frequently unmerchantable trees or slash were left behind. Over time, silviculturists noticed that residual trees provided a seed source and, given a good seed year and/or time, could result in successful natural regeneration—especially if grazing damage was prevented. Based on these observations, silviculturists began to purposefully select trees of the desired species, form, and vigor to serve as seed trees (Hausrath 1982). Since the focus of this practice was to facilitate natural regeneration, seed trees were removed after tree seedlings became established. However, because of the unpredictability of natural regeneration and the improvements in artificial regeneration techniques, the global application of the seed-tree method remained quite limited (Mantel 1990).

Strip Clearcutting

Foresters developed an alternative to seed-tree cutting for providing natural regeneration from seed. Observing natural regeneration in clearcuts, they noticed that edge or border (*Saum*) trees acted as a seed source for areas near the clearcut edge. This phenomenon was the basis for strip clearcutting. Instead of dealing with logistical complications of the seed-tree system (more difficult layout, multiple harvesting operations), harvesting units were laid out as long, narrow strips. Initially, strip clearcutting often caused windthrow problems in adjacent stands. In recognition of this, harvested strips were then oriented according to the prevailing wind direction in wind-exposed areas (Hausrath 1982). Widespread implementation of strip clearcutting required landscape considerations (to prevent wind damage and facilitate harvesting operations). Cutting units were shaped and spatially and temporally arranged in the landscape to minimize the exposure of forest edges to high-intensity winds (Mantel 1990).

Shelterwood System

The shelterwood system originated under quite different circumstances than either the clearcut system or the seed-tree system. Shelterwood systems started in lower-elevation, hardwood-dominated forests. Here,

Table 1.1. Factors in the Development of the Shelterwood System

	Beech	*Oak/Beech*	*Silver Fir*
Shade tolerance of regenerating species	Yes	No	Yes
Regeneration benefits from cover (frost/shade)	Yes	Yes	Yes
Homogenous stand conditions	Yes	No	No
Longevity and health of residual trees	Yes	Yes	No
Cover reduces understory competition	Yes	Yes	Yes

This table displays the assessment of selected factors that were either influential (Yes) or not relevant (No) in the development of the shelterwood silvicultural system in beech-dominated stands, mixed stands of oak and beech, and stands dominated by European silver fir in central Europe. For beech-dominated stands, all factors were relevant and the shelterwood system is still a common practice today. The lack of relevance of some factors for mixed oak/beech and silver fir–dominated forests indicates the dominant reason why the shelterwood system in these forests was not successful and was subsequently abandoned.

coppice with standards was a long-standing practice, but problems started to emerge after multiple cuttings. The sprouting vigor of repeatedly cut stumps declined. In addition, growth of sprouts was often reduced due to a dense overstory of the standards, especially in stands with large beech. Silviculturists noticed, through careful observation, that the tree overstory provided frost protection and suppressed development of competing vegetation (tab. 1.1), thus allowing new tree germinants to become established (Hausrath 1982). To take advantage of these conditions, the provision of shelter was accommodated through harvesting operations that left the largest, most vigorous trees.

The shelter trees were specifically selected to provide seed sources and protection for the regeneration. It became quickly apparent that seeds could germinate under fairly limited light conditions, but that seedlings required more light for continuous growth and survival. In response to these observations, the dense overstories were thinned after regeneration was established, and the silvicultural system was labeled *Hessischer* shelterwood after its region of origin in Germany (Hausrath 1982; Mantel 1990).

The shelterwood method during early development focused on providing optimal conditions for regeneration. The development of the regeneration dictated management and harvest schedules of overstory trees. This narrow focus did not account for local "necessities or needs," as it interfered with the demand for a predictable, continuous wood supply. To allow more flexibility in harvesting and accommodate market demands, Hartig (1791) and other silviculturists suggested and implemented the shelterwood system using multiple, repeated cuttings, including preparatory cuts and multiple removal cuts (Hausrath 1982). The shelterwood system provided a good compromise between economic and ecological constraints and also permitted variations in cutting intensity to reflect the density and growth of regeneration. This flexibility is one reason why shelterwoods are still common in many areas today. Limitations of the methods, however, became apparent when it was attempted in European silver fir (*Abies alba*) or mixed beech/oak forests. In those forests, the system consistently failed for reasons such as wind damage, decay, slow response of older coniferous shelter trees (in silver fir stands), and the inability to accommodate regeneration of multiple species with different shade tolerances (in beech/oak forests) (Hausrath 1982).

Femelschlag Systems

The *Femelschlag*, a type of patch or irregular shelterwood, gained widespread interest in forest types where shelterwood cuttings were attempted with limited success (see tab. 1.1). The development of the *Femelschlag* system was an important milestone, because it signaled a switch from managing regeneration at the stand level to more flexible applications that were adapted to conditions at smaller spatial scales (Vanselow 1963). Because of concerns about the stability of the largest trees in conifer-dominated forests, silviculturists harvested large, valuable trees first to capture their economic value. The system required smaller and/or younger and vigorous shelter trees to be left, as they were able to respond well to release, and therefore did not require immediate removal once regeneration had been established. Also, conifer stands were typically composed of patches with different tree densities and sizes, which could not easily be accommodated in shelterwood cuttings.

Working at smaller scales, silviculturists learned to open up the overstory in small patches through subsequent repeated cuttings. These openings were following a pre-described or regular pattern, but their location and treatment were determined by the local conditions. This system allowed for multiple species of different shade tolerance to regenerate over a few decades. *Femelschlag* was labeled after the region of origin (Baden, Germany: *Badischer Femelschlag*; Hausrath 1982; Hasel 1985; Mantel 1990). Applications of this system in other regions—for example, in areas with steeper slopes where multiple harvesting operations were not feasible—led to problems. When removal cuts were fewer and more aggressive, they failed to achieve the variety of conditions suitable for regeneration of multiple species and frequently resulted in need for planting (Vanselow 1963).

The importance of adapting silvicultural systems to specific local conditions was obvious when the *Badischer Femelschlag*, as described above, was compared with the *Bayerischer Femelschlag* (Bavarian *Femelschlag*, labeled after the region of its origin: Bavaria, Germany). Both were initiated from the unsuccessful applications of the shelterwood system and both worked with local spatial and temporal variability (Gayer 1886). In Bavaria, however, the reasons for shelterwood cutting's failures were mainly attributed to its inability to accommodate the high light requirements of oak seedlings while at the same time maintaining a sufficient closed canopy to reduce competing vegetation and provide conditions suitable for regeneration of beech (Hausrath 1982). As a solution, the *Bayerischer Femelschlag* opened up patches more aggressively to allow oaks to regenerate first. In contrast to the *Badischer Femelschlag*, which regulated regeneration progress through multiple removal cuts of the overstory above the regeneration, the *Bayerischer Femelschlag* emphasized the influence from edge trees around the regenerating patches. It focused further removal cuttings on these edges and opened up the overstory by more aggressively expanding the opening (Vanselow 1963). In both cases, however, cuttings were timed to accommodate the respective shade tolerances of the regenerating species, in most cases oak and beech. Thus, while on the surface both *Femelschlags* appeared quite similar, they should not be lumped in teaching and application (Hausrath 1982; Hasel 1984; Mayer 1984; Mantel 1990). A closer look quickly exhibits subtle differences that are a direct response to the local conditions described above.

Single-Tree and Group Selection

The single-tree selection method (also labeled *Plenterwald* in central Europe) originated under specific local conditions. Its origin is of special interest, as it is often cited as an alternative in regions where clearcutting is undesirable (e.g., Benecke 1996). A closer look at the history of development of the two selection systems suggests caution when advocating broader application of either system. Initially, *Plenterung* was synonymous with high-grading, the unregulated harvest of the most valuable trees with little consideration for regeneration or future stand productivity (Vanselow 1963; Mantel 1990). The widespread application of *Plenterung* had devastating effects on forests, leading to widespread areas with no or poor-quality trees (Vanselow 1963). Consequently, between the fifteenth and eighteenth centuries, regulations that prohibited *Plenterung* were put in place in many regions (Vanselow 1963).

Starting in the nineteenth century, the practice developed into a formal silvicultural system. The move toward single-tree selection in several forest regions in central Europe was aided by regulations that outlawed clearcutting (e.g., the 1810 forest law in Baden, Germany) mainly to avoid erosion and/or provide permanent avalanche protection on steep slopes (Mantel 1990). A second boost came early in the twentieth century when Biolley (1920) developed the *control method* and clearcutting was outlawed in Switzerland (1902 for designated protection forests, 1922 for public and private commercial timberlands).

The ability of single-tree selection to continuously provide a variety of forest products on small holdings fit the ownership structure in many regions in central Europe. Owners of small farms not only valued that these forests could act as a "savings account," but also valued the frequency of timber harvests for the constant cash-flow they provided. Furthermore, the cuttings could be done during the less busy winter months, which allowed implementation of the frequent intensive forest management practices necessary to maintain the uneven-aged stand structure (Mantel 1990). It is no coincidence that the prime showcases for single-tree selection (e.g., Emmental in Switzerland or the southern Black Forest in Germany) are composed of shade-tolerant tree species, such as European silver fir. The regeneration of mixed European silver fir and Norway spruce (*Picea abies*) stands required larger openings leading to development of the group-selection system.

The special combination of shade-tolerant trees, legal constraints, ownership patterns, labor availability, and market considerations that allowed the single-tree selection systems to be applied successfully was limited to a few regions in central Europe. Consequently, the use of single- or group-selection systems has been limited until recently. With the current interest in applying these systems broadly to achieve environmental goals, it is important to remind silviculturists that the underlying philosophy of single-tree and group-selection methods was the desire to achieve the highest harvest levels possible given the social and tree autoecological constraints of a particular region. Factors such as preserving aesthetics, mimicking effects of natural gap dynamics, or economic considerations were not considered at that time (Mantel 1990). The history of the development of these systems needs to be closely evaluated before recommending widespread application of single- or group-selection systems in various parts of the world.

Adoption versus Adaptation

Throughout the history of silviculture, silviculturists have struggled to weigh the benefits and costs of applying true-and-tried practices versus developing new tools specific to local conditions. An appreciation of this conflict is important for our current understanding of silviculture, and two aspects deserve to be mentioned here. First, when faced with the challenge of managing forests in new regions, silviculturists looked to experiences in other regions as a starting point for their efforts. The beginning of the forestry profession in North America around 1900 provides an example for the adoption of silvicultural practices from other regions (central Europe), and issues related to this adoption are discussed at length earlier (Hawley 1921; Spurr 1956; Weetman 1996; Graham and Jain 2004). Second, silviculturists who had developed a successful local practice were often convinced of the generality of their findings and consequently encouraged or enforced the widespread use of this practice beyond the conditions where it was developed (Hausrath 1982; Mantel 1990).

Lack of scientific understanding about the importance of local site adaptations and/or the desire of silviculturists for self-promotion and regional or national recognition were often behind the generalization of local practices. A lack of ecological understanding was likely responsible for the application of silvicultural systems to different species within a region (see discussion about shelterwood and *Femelschlag* above, tab. 1.1). Applications of inappropriate silvicultural practices were less quickly discarded when they were forced upon a region through regulatory means, commonly by dominant or well-recognized silviculturists or administrators. In cases where silviculturists had a personal stake, their status and reputation could delay abandonment of unsuccessful practices for decades after problems were identified.

Examples of such behavior go beyond silviculture and include Georg Hartig's approach to forest inventory and planning (Hartig 1795). Another prominent example of a dogmatic attempt to apply a single silvicultural system to a whole region was Christoph Wagner's version of edge shelterwood (*Blendersaumschlag*) (Wagner 1912). Wagner developed the system while employed by a private landowner in southwestern Germany. Later, after becoming state forester of Württemberg, he encouraged a regulation that required implementation of the system on all state forest lands (Vanselow 1963; Mantel 1990). The fact that cutting patterns in edge shelterwoods were based solely on windthrow concerns and ignored the local ecological conditions and tree species composition was recognized by the forestry profession, and the edge shelterwood (*Blendersaumschlag*) was quickly abandoned (Mayer 1984; Mantel 1990).

Even more forceful were attempts by K. Phillip to apply his "wedge shelterwood" (*Keilschirmschagverfahren*) to forests in the entire state of Baden, Germany (Mayer 1984). First, he used his position as a state forester to enforce the wedge shelterwood system on all state lands. He further drafted a state law that extended the wedge shelterwood system to all forestland in the entire state. The law was heard and discussed in the parliament, but defeated in a vote by the legislature (Mayer 1984; Mantel 1990). As can be expected, these and other attempts to apply silvicultural systems and practices broadly were abandoned after the personalities that championed the ideas lost their status.

Other attempts to apply silvicultural practices globally have been quite successful. The clearcutting system has been applied broadly in North American forests mainly because of economic reasons. The wider application of other silvicultural systems was often driven by similarity in ecological constraints. For example, shelter trees benefit regenerating seedlings in areas with frost concerns (e.g., in Sweden and the Black Forest, see Langvall and Orlander 2001 and Burschel and Huss 1997, respectively) and in areas with high summer temperatures (e.g., in southwestern Oregon, see Childs and Flint 1987).

Integration of Scientific Advancement into Silviculture Teachings

Silviculture, by its very nature, is an integrative field that synthesizes information from other scientific disciplines (Nyland 2002). As such, it is strongly influenced by the advancement of knowledge in other disciplines. Early forestry books listed silviculture as the central discipline of forestry (du Monceau 1766; Hartig 1791) and covered ideas and knowledge about growth and yield, inventory, and economics. These books focused on management activities, with little discussion of scientific underpinnings (Cotta 1816; Pfeil 1851).

In both English and German, the name "silviculture" (*Waldbau*) indicates that the discipline is analogous to agriculture (*Feldbau*). This choice of name for the new discipline of silviculture was a direct reflection of Cotta's opinion that forest management was equivalent to agricultural cropping. By the middle of the nineteenth century, silviculture writings were dominated by the influence of inventory and planning systems. It was not until Gayer (1880, 1886) that the importance of ecological considerations for silviculture was recognized and acknowledged (Mayr 1909). The scientific understanding of ecological processes, however, was quite limited at the time and Gayer (1880) relied heavily on his personal experience and general understanding of forest ecosystems to cover this topic. For example, Mayr (1909) pointed out that the forest inventory pushed stand layouts toward "geometric shapes." He credited the importance of economic considerations over ecological conditions, which would suggest smaller-scale management units.

Early in the twentieth century, silvicultural books began to formally acknowledge scientific findings as a basis for silvicultural decision making. For example, Mayr (1909) first presented scientific data in his silviculture book. Morosov (1920) included the first treatment of site and stand-type classifications. Shortly thereafter, scientific progress in subjects like botany, climatology, and soil science was addressed and the importance of site characteristics on growth of forest trees was recognized (Rubner and Leiningen-Westerburg 1925). A major shift was signaled by Dengler (1930), who stressed ecology as a basis for silviculture. Dengler relied heavily on increased scientific understanding of the influence of site conditions on forest development and other ecological relationships for making silvicultural decisions (Mayer 1984). By the middle of the twentieth century, the view of forests as ecosystems and the interrelationship of ecological components were well integrated within the silvicultural literature (e.g., Köstler 1949; Leibundgut 1951). These writings emphasized that silviculture should not be simply viewed as long-rotation agriculture but should base its decisions on an understanding of plant communities and ecosystem dynamics, a view that is still widely held in central Europe today (Mayr 1984; Burschel and Huss 1997).

A viewpoint that has received repeated attention in forestry teaching in the twentieth century is the development of the "permanent forest" (*Dauerwald*) movement. The permanent forest movement is an example of how an underlying philosophy can influence management practices. Early supporters of the *Dauerwald* viewed forest ecosystems as a single organism (Möller 1923). On the assumption that an organism could be preserved only if all processes were maintained at all times, this movement favored a permanent forest cover. This view was not compatible with the mainstream thinking of silviculturists and led to discussions about the suitability and profitability of this approach that are ongoing (Wiedemann 1925; Jakobsen 2001).

Throughout much of the later twentieth century, close-to-nature forestry cover was practiced by only a few landowners, and the practitioners did not receive much attention from mainstream research and educational institutions. Starting in the 1990s, with the emergence of a more eco-centric view of forests and forestry, interest in permanent forest, or continuous cover forestry, has been revived by close-to-nature forestry movements in Europe (Thomasius 1999; Jacobsen 2001; Pommerening and Murphy 2004) and to a certain extent by ecosystem management, retention harvest, and ecoforestry in Canada and the United States (Kohm and Franklin 1997), Japan (Fujimori 2001), and New Zealand (Benecke 1996). The rise of close-to-nature movements in Europe was not a direct response to new research findings. Instead, in Germany it was mainly carried by practitioners looking for alternatives to conventional forestry practices, who were loosely organized in the Arbeitsgemeinschaft Naturgemäße Waldwirtschaft (ANW, Association of close-to-nature forestry, Hatzfeld 1995). The devastating windstorms in central Europe during the spring of 1990 became a turning point for the movement in terms of widespread interest. Pro-Silva evolved as a European network in 1999 (www.prosilvaeurope.org) and developed formal principles that include natural regeneration, continuous forest cover of mixed-species stands, and allowing natural processes in all aspects of silviculture (Pommerening and Murphy 2004).

A parallel historical assessment points out that early North American literature started by describing silvicultural systems developed and established in Europe (Hawley 1921). Among other reasons, this may simply reflect the very limited understanding of the ecology of forest ecosystems at the time. For the next decades, new textbooks (e.g., Toumey 1928; Westveld 1939; editions of Hawley's 1921 book) shifted incrementally to include an emphasis on scientific ecological understanding. For example, Daniel et al. (1979) emphasized the scientific basis for tree and stand growth.

During these decades, the study of how stand structures change over time developed into a new subdiscipline, stand dynamics (Oliver and Larson 1996), that now greatly influences silvicultural writings (e.g., Smith et al. 1997). The development of stand dynamics also highlighted silviculture becoming a global discipline, as it was the first major silvicultural concept that was initiated in North America and transferred back into the European literature (Otto 1994). Other examples of this phenomenon include the concept of "ecosystem management" or its derivatives (Kohm and Franklin 1997; Puettmann 2000; Lindenmayer and Franklin 2002), which stress the maintenance of the full array of forest values and functions at the landscape level.

Practices associated with ecosystem management are now gaining attention in Europe, such as retention harvest, in which structural elements like large live trees, dead standing trees, or logs on the ground are retained after harvest as biological legacies (e.g., Vanha-Majamaa and Jalonen 2001). The notion of emulating natural disturbance patterns in silvicultural decision making (Bergeron et al. 1999b; Burton et al. 2003; Perera et al. 2004) was initiated in North America and is now starting to gain attention worldwide (Kuuluvainen 2002; Lindenmayer and Franklin 2002; Franklin and Lindenmayer 2003). As new trends spread to other regions and countries, they will be faced with challenges similar to those faced by the traditional European silvicultural systems when introduced into North America. For example, the notion of natural range of variability or implications of natural disturbance regimes has a different meaning in areas with a long history of human influence on forest ecosystems (Puettmann and Ammer 2007). How these new concepts and evolving practices influence silviculture writings and the evolution of silviculture as a discipline in the various regions of the world is yet to be seen.

Conclusion

The history of silviculture teaches us to view silvicultural practices from the perspective of the assumptions and external conditions that produced them. The link between silvicultural practices and external factors is tight, yet often receives little attention by silvicultural researchers and instructors. Our review of the history of silviculture should eliminate any expectations that the silvicultural systems developed around 200 years ago, and influenced by wood famines and the onset of liberal economic thinking, would automatically be suitable to handle present-day and future challenges. Nor should silviculturists expect practices developed for specific regions and/or species mixtures in Europe to be capable of handling forestry challenges in other parts of the world, such as North and South America, Africa, or Asia. That said, there is also a lot that can be learned from past experience; after all, silviculture has worked well in many places where forestry had a clearly and narrowly defined objective. Our historical review of silviculture suggests the discipline has a long and successful history of adapting practices in response to altered social, economic, or ecological conditions.

2

Silviculture

Challenging Traditions

The discipline of silviculture is the management and study of forests to produce desired attributes and products. Silviculture has strong traditions that have been developed, articulated, and refined over several centuries (chap. 1). Throughout this time, the objective of most landowners, and therefore of most silvicultural activities, has been the efficient production of wood for timber or other wood-based commodities. Accordingly, silviculturists have successfully focused on developing practices to efficiently regenerate forests and increase wood production and quality.

Although there has been, and continues to be, a strong emphasis on wood production in silviculture, the discipline should not be considered a homogeneous field. The management of seminatural woodlands and protection forests are also aspects of silviculture. Throughout history, silvicultural principles have been used to manage forests to promote wildlife habitats, to ensure hunting opportunities, to provide reliable sources of clean water, to protect settlements from snow or rock avalanches, and to establish and maintain tranquil forest settings.

Silvicultural practices, regardless of management objective, aim to control the establishment, composition, structure, growth, and role of trees within managed forests. Preferred tree species are established through natural regeneration, direct seeding, or planting. Composition refers to the variety of tree species and their relative abundance. Structure comprises the internal characteristics of forests including tree crowns, vigor, diameter and height distributions, the abundance and types of dead trees (snags), the presence of wood on the ground, and understory vegetation. Silviculturists manage tree growth and quality by manipulating tree species composition and density, by removing other competing vegetation, and by improving site productivity. They manage habitats by retaining or promoting specific forest structures such as snags and old large trees.

Silvicultural activities are implemented through a series of individual practices (e.g., site preparation, promoting natural regeneration, planting, fertilization, thinning, and final harvest of individual trees or stands based on diameter or age; see Hawley and Smith 1972; Daniel et al. 1979; Burschel and Huss 1997; Smith et al. 1997; Fujimori 2001; Nyland 2002) that promote the desired species and structural characteristics within and among managed areas in a forested landscape. Individual silvicultural practices are integrated into a silvicultural system, which can be viewed as a larger program of activities aimed at achieving desired tree composition and growth objectives (see chap. 1). Probably the single greatest defining characteristic of the discipline of silviculture is the concept of silvicultural systems and their application in the management of forests (Troup 1928; Matthews 1989; Mantel 1990). While individual practices have changed over the years based on better understanding of their impacts or new technologies, the suite of even- and uneven-aged silvicultural systems formalized in central Europe in the nineteenth century are still being applied today in forested regions throughout the world with surprisingly few modifications. As a result, silviculture across the globe has a common origin. The basic structure and principles of the discipline are often considered to be independent of local conditions (Hawley and Smith 1972; Burschel and Huss 1997; Fujimori 2001; Nyland 2002).

The discipline of silviculture can be best understood by examining five core principles that have formed the basic foundation of silvicultural thinking, study, and practice: (1) a strong focus on trees to the exclusion of other plants, animals, and ecosystem processes, where these are not relevant to the task of growing trees; (2) conceptualizing stands of trees as uniform management units; (3) applying an agricultural approach to silvicultural research, especially the search for best treatments that emphasize uniform tree species composition and structure; (4) the scale-independent view of silvicultural practices; and (5) a strong desire for predictable outcomes.

The core principles are focused on the most dominant objective of silviculture to optimize the quantity and quality of wood products. They have guided silvicultural practice globally and remain a strong influence in contemporary silvicultural thinking and practice. While exceptions clearly exist, we believe that silviculture as a discipline is strongly influenced by entrained thinking and tradition, and that insights can be gained by all silviculturists in reviewing the set of core principles in the context of their influence on addressing present-day issues.

A Dominant Focus on Trees

The development of natural sciences, including silviculture in the seventeenth and eighteenth centuries, reflected the writings and beliefs about nature of the principal philosophers and scientists of the time (e.g., Descartes, Newton, and Kant). Rational thinking and Newtonian mathematics implied that nature, and therefore forests, were driven by universal laws. It was considered man's obligation to bring order to nature. This rational view of the natural world was heavily influenced by Newtonian mathematics, which relied on simplification and linear relationships (Hampe 2003). While it is unlikely that many silviculturists read Newton's writings, the philosophical view of nature still influenced their work. For example, straight lines or sharp edges were perceived as superior by Newton, likely influencing the linearity and regularity of early silvicultural operations, especially those in Europe aimed at reforesting highly degraded forests to mitigate the wood famine (see chap. 1).

Early silviculturists managing for wood production believed they were enhancing the ultimate goal of nature by taming nature; that is, transforming degraded woodlands or natural forests into more orderly arrangements of desired tree species with balanced age classes (see normal forest discussion in chap. 1). To tame nature, silviculturists developed a suite of practices that centered on controlling biotic and abiotic conditions to reliably enhance the performance of the tree species with the most desirable growth and wood properties (for a more thorough discussion of these practices, consult silvicultural textbooks).

Most early silvicultural practices aimed to make forests conform to this worldview. For example, unproductive sites and dead trees were seen as a waste and thus were restored to productivity by drainage or fertilization or removed in harvesting operations, respectively. Despite many

Taming Nature: The philosophical view that the "messy" natural forest needs to be transformed into a forest that is improved and superior has a long history in silviculture. Expressions in French (*il faut éduquer la forêt*) and German (*Walderziehung*) implied that the natural forest needed to be "trained" or "educated." This translated into simplifying forest structures and uniform conditions. The concept that managed forests are better than natural forests in achieving ownership objectives is still evident in contemporary silvicultural thinking: "in silviculture, natural processes are deliberately guided to produce forests that are more useful than those of nature, and to do so in less time" (Smith et al. 1997, 5).

subsequent changes, most notably in our ecological understanding of forest functions, this worldview remains pervasive in contemporary silvicultural thinking and practice. Especially in plantation management but also, in different dimensions and to a different degree, in management of seminatural woodlands, the "obligation to bring order" combined with economic efficiency resulted in uniformity of forest practices and simplified forest structures. The desire for order and simplification is even evident in intensively managed present-day, uneven-aged forests.

The most visual evidence of the silvicultural emphasis on regularity and evenness is the control of tree density and spacing in managed even- and uneven-aged forests. In plantation management, trees are planted in square or rectangular spatial patterns. In natural stands with dense natural regeneration, regular tree spacing is achieved through thinning. Often, the first thinning entry is focused on providing regular, optimal growing conditions, rather than a direct economic return. It is thus labeled precommercial thinning or spacing because trees are usually too small to be sold profitably. Commercial thinning takes place in older natural or managed stands where the cut trees can be sold. In managed uneven-aged forests the number of trees allowed in various diameter classes and the size at which the largest trees are cut is controlled to promote maximum growth onto the selected trees. The major aim of the control of tree density and spacing in managed even- and uneven-aged forests is to focus the full growth potential of a site to a limited number of desired trees and thus maximize economic gain.

In efforts to control and improve on nature, genetic improvement programs were developed to select seeds for regeneration from parent trees with superior growth and wood quality. Plantations in New Zealand, Chile, and Argentina that were established with an extremely narrow focus on specific wood products provide the most remarkable examples of impacts of silvicultural practices aimed at maximizing wood production. Displaying a striking difference from native forests, monospecific plantations in these regions are even-aged, with evenly spaced trees of similar size and form. Furthermore, these plantations are typically composed of tree species that are not native to the area.

All practices described above, to lesser and greater degrees, aim to develop an ideal forest that is composed mostly (or preferably only) of vigorously growing, healthy trees of high wood quality, most commonly in single-species even-aged stands, but also in mixed-species or uneven-aged stands. Desired trees are now often referred to as "crop trees," a term that implies trees can be managed like crops in an agricultural field (Cotta 1816).

The emphasis on controlling species composition and spacing to enhance tree productivity and value remains an influential feature in the discipline of silviculture, as encapsulated in a quote from Smith et al. (1997, 4): "silviculture for timber production is the most intricate kind because the species and quality of trees are of greater concern than they would be with other forest uses." This view has many advantages, one of them being that the successes of silvicultural practices were quantifiable by measuring the quantity and quality of trees.

The management goal of timber production and the associated emphasis on trees also provides a clear picture of what a successful, well-managed forest should look like, one that efficiently provides homogeneous, high-quality timber. Consequently, regions that practiced intensive silviculture following this approach gained reputations as examples of good forestry and became the subject of many field trips and excursions. For example, the intensively managed Scandinavian or New Zealand plantations have long been considered showcases of successful industrialized forestry operations. Alternatively, forests managed intensively by uneven-aged single-tree selection (e.g., *jardinage* or *Plenterwald*) (Matthews 1989), and more recently by "close-to-nature" approaches, have drawn visitors as showcases of successful silviculture in central Europe (Jakobsen 2001; Pommerening and Murphy 2004).

Because trees are long-lived organisms, silviculturists have had a longer-standing familiarity with the concept of sustained yield and sustainability than have experts in most other disciplines (Peng 2000). The focus of silviculturists on trees, however, also limited the scope of their interest in sustainability (Morgenstern 2007). The sustainability principle can be traced back to von Carlowitz (1713), who was interested in ensuring a continuously high wood supply for mining needs. Of course, in some areas in central Europe, sustainability of hunting opportunities for landowning nobility was another early concern to silviculturists. The vast majority of silviculturists, however, have come to equate the sustainability of forests with the sustained yield of timber (Morgenstern 2007). One in-

Sustained Yield and Sustainability: Trees and forests are renewable resources, so it is appropriate to discuss sustainability, which is the ability to maintain something undiminished over time (Lélé and Norgaard 1996). Sustained yield assumes that any tree species or community of tree species produce each year a harvestable surplus that can be harvested so as to maintain the capital and the productivity of the forest (Larkin 1977). The meaning of sustained yield, as applied to the management of trees for timber production or deer for hunting, and the concept of sustainability of forest ecosystems are distinct though related concepts (Hilborn et al. 1995). Sustainability encompasses a wider array of resources and values and has ecological, economic, and social dimensions (Levin 1993).

herent feature of such a strong management focus on trees was the acquired belief that other characteristics of the forest ecosystem would benefit or at least not be harmed by such management activities. This is reflected in the statement that "what is good for the trees is good for the forest." The implication was that forests managed for timber production would also automatically provide all other forest values and functions. This continued to be a strongly held belief until recently (Pretzsch 2005).

Because silviculturists have tended to view forest ecosystems through a tree-focused lens, other components of forest ecosystems were often considered only in terms of their impact on individual tree survival and growth. For example, herbs, shrubs, and trees other than the desired tree species were not managed in relation to their potential contributions to nutrient cycling (Attiwill and Adams 1993) or wildlife habitat (Hunter 1990). Instead, the major interest of silviculture in dealing with these ecosystem components was to limit their competition with crop trees (Wagner et al. 2006). Especially in plantation management, the focal point of silvicultural attention on other forest plants was their reduction or elimination (Walstad and Kuch 1987; Thompson and Pitt 2003; Wagner 2005). Silviculturists have generally evaluated ecosystem processes only in the context of their management goals. For example, interest in mycorrhizae fungi was focused on the potential beneficial effects of the fungi to seedling establishment and tree growth. Whether harvesting altered fungal communities or how the removal of competing vegetation impacted fungi and subsequent ecosystem function generally received little or no attention by silviculturists. With the wider range of management objectives, especially on public forests, the tree-focused nature of silviculture is undergoing a recent change (see examples in chap. 4).

Natural disturbance agents in forests were also viewed and managed in the context of their impact on tree and stand productivity. Decay fungi and insects were seen as damaging agents and discussed under the topic of pest control in silviculture classes and writings. Until recently (see chap. 4), disturbances such as fire or windthrow were mainly assessed in terms of their damage to trees and stands rather than as relates to their role in succession and ecosystem function. Unplanned disturbances were often labeled as catastrophes, and considerable silvicultural efforts were aimed at preventing or minimizing impacts of disturbance to ensure a predictable high level of tree and stand growth.

The idiom "can't see the forest for the trees" implies an excessive concern with detail resulting in a lack of understanding of the larger situation. As we learn more about the complexity of forest ecosystems (see chap. 3), we'll see that the idiom can be applied more literally to characterize the discrepancy between the emphasis on trees by traditional silviculture and our current understanding of how much more there is to a forest than just its trees.

Management of Stands as Uniform Entities

As silviculture evolved into a well-established discipline, the notion that forests should be managed on a stand-by-stand basis emerged as a key concept (Smith et al. 1997; Helms 1998). A stand is the most basic unit of management in forestry, consisting of a contiguous group of trees sufficiently uniform in age-class distribution, composition, and structure, and growing on a site of sufficiently uniform quality to be a distinguishable unit (Helms 1998). Stand management has resulted in efficient planning and inventory procedures, and the prevalence of managing homogeneous units has considerably influenced silvicultural thinking and views of forest ecosystems.

The delineation of stands in unmanaged forests is determined by landscape topography and prior disturbance events. Disturbances as determinants of stand size and boundaries deserve special attention. First, disturbances are fundamental to the development of structure and composition (attributes that help identify a stand) and maintenance of forest health and productivity (Oliver 1981; Attiwill 1994). Second, regional natural disturbance regimes have frequently been used to justify stand sizes and harvesting patterns. In most regions, however, natural disturbances in forests vary spatially and temporally from frequent small-scale, low-intensity, gap-forming disturbances operating at the level of individual trees to larger-scale, high-intensity events that affect large areas (Spies et al. 1990; Frelich 2002; Johnson and Miyanishi 2007). Thus, while both small- and large-scale disturbances are common in many forests, identifiable stands result mainly from medium- to large-scale disturbance events such as fires, windthrow, or severe insect infestations that kill most trees and result in relatively uniform regrowth. The preoccupation with delineating the external boundaries of stands based on large-scale mortality events allowed silviculturists to overlook small-scale within-stand variability as an alternative means of characterizing stands.

In regions where silvicultural management started with regenerating degraded areas or harvesting of natural forests, stands were typically

Figure 2.1. Example of a forest ownership with stands of Norway spruce and European beech in Sauerland, Germany. Note that stands are of small size and stand boundaries are obvious. Picture credit: Irene Breil.

delineated by logistical constraints. Harvest unit layout usually reflected concentrations of trees that were of greatest interest to loggers and topographic conditions. Stand size could vary considerably depending on physical, social, and historical constraints. The size and shape of the area harvested was often determined by the requirements of logging equipment or property boundaries. More recently, government regulations in most jurisdictions have put some limits on the size of harvesting units, which adds another element to how stands are delimited. In many regions of the world, stand boundaries were established centuries ago and subsequent silvicultural practices have ensured easy identification of the individual stands in the landscape (fig. 2.1).

Management intensity, ownership, and land tenure pattern also influence stand size. The size of individual stands can be quite small in the intensively managed, privately owned boreal forests of Finland (typically ranging from 0.5 hectare to 50 hectares) but are much larger in the more extensively managed publicly owned boreal forests of Canada and Russia (one hundred to several thousand hectares). Stand sizes in areas with longer management history tend to be smaller, reflecting early logging constraints and historical ownership patterns. For example, all private forestland in Croatia and Poland is in parcels of less than 5 hectares (Food and Agriculture Organization 1997) compared to 10 to 11 hectares average stand size on industrial, public, and tribal lands in Minnesota, United States (Puettmann and Ek 1999), and approximately 22 hectares on land owned by the forest industry in the Pacific Northwest of the United States (Briggs and Trobaugh 2001).

Probably the most influential aspect of stand management on silvicultural thinking is the traditional use by silviculturists of tree-based stand descriptors such as stems per hectare; tree diameter and height; current, periodic, or mean annual increment; basal area; merchantable or total volume; diameter distributions; and the "q-factor." Most of these descriptors are based on the assumption of underlying normal distribution, with one exception. The q-factor has a special place in silvicultural history and has been used to prescribe the desired diameter distribution of stands managed by the selection system (see chap. 1). The q-factor, first proposed by the French forester de Liocourt (1898), is an indicator of

Stand Descriptors: Current annual increment is the amount of wood that stands add in any given year, whereas periodic annual increment is for some fixed period of time, usually five or ten years. Mean annual increment is the average amount of wood accumulated each year over the full life of a stand. This is a key value for determining a sustainable harvest level. Basal area per unit of land (square meters per hectare) is a measure of the cross-sectional area of tree trunks in a stand. It is easily measured using a prism and is a common means to describe stands. Total volume is the gross wood/stem volume of all trees in a stand (cubic meters per hectare), whereas merchantable volume includes only tree stems above a minimum size threshold.

the ratio of the number of large trees to the number of small trees in a stand. Mathematically, it is reflected in the steepness of the negative exponential (reverse J-shaped) diameter distribution common to uneven-aged stands (see also figs. 5.4 and 5.6).

Descriptors are usually averaged over the whole area of a managed stand. These averages are commonly used to describe stand structures and for planning timber management activities. Obviously, basing decisions on average stand conditions implies that stands are sufficiently homogeneous to be properly represented by an average value. Similar assumptions of homogeneity within stands are also inherent in stand descriptors that describe growing conditions. For example, the growth potential of a stand is frequently represented by the site index of the desirable tree species. Site index is a common, useful, and widely used measure in silviculture. It is also an example of the deeply entrenched focus on homogeneous stands. Site index utilizes trees as a bio-indicator of the potential productivity of a site and requires those trees to have grown without overtopping or any significant reduction of height growth. This

---

Site Index: The average height of the dominant and codominant trees at a specified age (SI_{50} = height at age fifty). Tree age is often determined 1.3 meters above ground level, at "breast height." Site index is a tree-centered quantitative metric that is used to express site productivity. It is based on the assertion that height growth is independent of crowding and thus reflects inherent site conditions. Since tree species have different growing requirements, the site index metric is species specific. Individual trees selected to determine site index are assumed to have grown without ever being overtopped by other trees.

---

limits its utility to uniform even-aged, single-species stands, and its use may thus implicitly encourage uniform stand management practices.

The focus on average stand descriptors with their inherent assumption of homogeneity has also become the standard method of describing silvicultural practices. Individual prescriptions for silvicultural practices like planting (or thinning) propose a certain number of stems per hectare to be planted (or retained in a thinning) within an allowable deviation, typically limited in contracts to 5 or 10 percent. Prescribed densities are used to calculate desired distance between trees based on square or triangular arrangements. These inter-tree distances are then evenly applied throughout the stand.

The notion that all areas within a stand are similar, or at least similar enough to be represented by a single number, worked well in managed even- and uneven-aged stands, such as the most intensively managed plantations or selection systems forests. At a broader level, the traditional use of average stand descriptors has trained silviculturists to think and view forest ecosystems in terms of uniform conditions that can be easily summarized by use of an average descriptor. On the other hand, the variability often associated with dynamic ecological systems like forests did not receive the same attention.

The desire to fit forest management into the industrial efficiency paradigm cannot be underestimated in its influence in promoting the stand concept and within-stand uniformity. Especially with the onset of larger mechanized machinery, silvicultural prescriptions needed to be designed to take full advantage of industrialized tools and methodologies. For example, the types of equipment used in harvesting operations often dictate minimum stand sizes for cost-effective operations. Maximum stand size is also limited by logistical constraints as the area that "can feasibly be treated in a relatively uniform manner" (Tappeiner et al. 2007, 34).

As sawmills became increasingly mechanized and streamlined, they typically limited their operation to a few selected tree species and more recently even to a narrow range of tree dimensions. This development pushed silviculturists to plant monocultures of desired tree species for efficient management. For example, planting monocultures avoids the cost of sorting logs by species to supply different sawmills. At the same time, it became more important to produce consistent log sizes and qualities, which required more uniform growing conditions within stands.

Stand-based management has gained worldwide acceptance and usage in forestry for planning and implementing silvicultural prescriptions and practices (e.g., Smith et al. 1997; Fujimori 2001; Röhrig et al. 2006). It has proved quite successful at achieving the goal of increased management efficiency and timber productivity. The stand concept, which is institutionalized as desired or good forestry practice, provides an example of how management practices that developed in response to economic and logistical constraints resulted in further homogenizing conditions within—by definition—already homogeneous stands.

Applying an Agricultural Approach to Silvicultural Research

The process by which the discipline of silviculture developed and adapted new practices and techniques has been very influential in how the discipline operates and how it views forested ecosystems. During the early development of silviculture, the refinement of individual silvicultural practices was based on long-term observation and local trials. The emergence of distinct silvicultural systems was not the result of a grand research effort to determine practices that could be implemented widely. Rather, silvicultural decisions, and therefore also silvicultural systems, were developed by refining local practices and experiences. Early local adaptation was not part of formal scientific experiments but rather an inherent part of application. This history is reflected in the intricate naming protocols employed by German foresters to describe site-specific modifications to even- and uneven-aged silvicultural systems (see chap. 1).

Contemporary silviculture is described as the *art* and *science* of managing forests (e.g., Smith et al. 1997). The art can be thought of as application of knowledge that is based on careful observation and long-term practice. Knowledge was gained from experience, which provided silviculturists the ability to match or modify existing successful practices to new management conditions. The art of silviculture become so ingrained in early practice that the word *Götterblick* (literally "God's insights"; often translated as "forester's belief") was used in the German language to describe when forest management decisions were based on experience, rather than on formal empirical relationships (Abetz and Klädtke 2000; Freise 2007). The strongest present-day example of the art of silviculture can be seen in the close-to-nature movement centered in Europe (Jakobsen 2001). Lacking a strong scientific database, this movement relies heavily on experience and a deep understanding of local conditions (Thomasius 1999; Jakobsen 2001).

The dependence on insight and experience for practice development resulted in a mind-set among silviculturists that relied heavily on tradition. For silviculturists to gain and utilize long-term experiences requires continuous employment in the same position or at least in the same region. In central Europe, during the nineteenth and twentieth centuries, it was quite common for silviculturists to manage the same forest throughout their career, in many cases for multiple decades. Furthermore, it was not uncommon for positions to be handed down within a family from one generation to the next. While careful observations and long job tenure ensured continuity of practices, it also resulted in silviculture becoming steeped in tradition.

Although this structure of the profession fostered long-term application of locally adapted practices, it did not encourage critical and innovative thinking (Brang 2007). Nor did the onset of formal education for silviculturists necessarily encourage innovative ideas and approaches. Instead, formal education led to greater regional (and later global) standardization of selected silvicultural practices. The emphasis on long-standing traditions is likely one reason why silviculture does not easily adjust to rapidly changing societal values. On the other hand, many silviculturists, correctly, still see these traditions as one of the strong assets of their profession. There are clearly trade-offs between using tried-and-true practices compared to switching to more short-term, flavor-of-the-day approaches.

Starting in the early part of the twentieth century, forest research stations were established and a scientific research approach began to be applied to silvicultural topics. The onset of formal scientific inquiries in forestry was closely linked to the development of experimental and statistical methods in agriculture, as "silviculture is to forestry as agronomy is to agriculture" (Smith et al. 1997, 3; see also Cotta 1816). In that context, silvicultural research borrowed heavily from agricultural research techniques, which were developed and employed to improve agronomic methods with the main purpose of maximizing farm crop yields.

Silvicultural research and associated educational efforts were strongly influenced by experimental and statistical advances. Most notably, contemporary statistical procedures for agronomy were developed and refined by the statistician Sir R. A. Fisher (1890–1962) at the Rothamsted Agricultural Experimental Station, England. Between 1919 and 1935, Fisher pioneered the design of experiments and analysis of variance (ANOVA). Silvicultural researchers were trained to use the classical agricultural experimental designs, including completely randomized, randomized block, Latin square, factorials, or variations such as split-plot designs (e.g., Petersen 1985). Silvicultural research today remains very much dominated by these statistical approaches and the use of designed experiments with all their strengths and limitations. Designed agricultural experiments and the associated analytical methods were originally developed to find techniques for increasing annual crop yield within agricultural fields. These experiments are therefore most appropriate when silviculturists are mainly interested in higher timber yields.

Agricultural experiments are designed to find a new practice, or best treatment, that optimizes a desired outcome, usually increased yield. An example of the application of agricultural experiments in a forestry setting is a study to determine whether exotic tree species will yield more than native tree species. Researchers would set up an experiment using one of the experimental designs developed by Fisher and test whether there is a statistically significant difference in stand yield between selected exotic species and the favored native species. The experiment is actually testing whether the null hypothesis (no difference exists in average yield calculated across all replications) among the tree species can be rejected. Null hypothesis testing to identify a "best treatment" is a cornerstone of the designed experiments used in agricultural and silvicultural science. If the null hypothesis is rejected, the new best species is expected to outperform others in operational plantations. Silviculturists

---

Null Hypothesis: Results of silvicultural experiments that rely on ANOVA are either a rejection of the null hypothesis or a failure to reject it. The researcher desires to prove that one of the new treatments will be superior (i.e., the null hypothesis will be rejected) and be suitable for broad application. Such experiments are not designed to assess the relative strength of observational support for alternate hypotheses. Despite considerable criticism of null hypothesis testing (Hilborn and Mangel 1997; Burnham and Anderson 2002; Johnson and Omland 2004; Stephens et al. 2005; Canham and Uriarte 2006), it remains the dominant statistical approach used in silviculture.

---

who utilize this new information will plant the new best species, until yet another best species can be found through experimentation.

Designed agricultural experiments that do not reject the null hypothesis are often considered a failure. First, they don't show progress—after all, the study did not lead to an improvement in management practices. Second, questions arise about whether limited sample size, high variability in study conditions, or other experimental constraints are responsible for the results. Third, studies that don't reject the null hypothesis are harder to publish (Csada et al. 1996). Researchers using null hypothesis testing are under pressure to find statistically significant results.

The use of designed experiments and null hypothesis testing by silvicultural researchers has strongly influenced the way field silviculturists view and implement silvicultural prescriptions. To fully appreciate the impacts of the agricultural research model on silvicultural practices, it is important to understand the suite of factors implicit in such designed experiments. These factors include null hypothesis testing, a defined suite of treatment factors, a limited set of treatment levels, the need for homogeneous treatment plots, the control of stochastic factors, and inference scope. We now discuss each of these in turn.

Thinning Studies: Probably one of the oldest types of silvicultural experiments. The recent controversy about thinning responses, initiated by Zeide (2001a) and followed up by letters and discussions in numerous settings, highlights limitations of agricultural research approaches. The discussion pointed out that the regression approach is not intrinsically different from ANOVA with all its assumptions and limitations. Zeide (2001b) suggests that after "centuries of research" we still do not understand the basic patterns of tree and stand responses to thinnings. He points out the "little utility" of empirical regression equations because they are "tied too closely to specific species, age, site, and other circumstances to be of general interest" and while being a "useful, heuristic tool . . . regressions are of little value to our knowledge." He proposes "conceptual generalizations based on the understanding of the involved processes" to avoid "going in circles."

In order to efficiently search for a new best treatment via null hypothesis testing, researchers can usually examine only a few treatment factors and/or treatment levels. The selection of the treatment factors and levels from an unlimited set of possible options can greatly influence the study conclusion. For example, a density study that compares stands with 100, 300, and 600 trees per hectare is more likely to find statistically significant differences than the same study setup with 200, 300, and 500 trees per hectare, and thus may come to different conclusions about impacts of density management. This shortcoming of null hypothesis testing becomes more limiting when issues are addressed that may entail interacting components, such as what factor, agent, or process is responsible for thinning responses or growth or mortality patterns. Furthermore, null hypothesis testing provides silviculturists with an implicit message that "scientific management" could simply imply picking the treatment from a limited set of possible options that performed best in experiments.

Types of Silvicultural Experiments: Most silvicultural studies fall into one of three broad groupings. First, and by far the most common, traditional agricultural experiments searching for a best treatment; for example, which thinning regime maximizes merchantable yield? Second, studies aimed at finding the best condition for a desired result; for example, seedbed requirements for good germination and early survival. Third, studies conducted across some type of gradient of conditions; for example, growth rates of juvenile trees under varying light levels or under different overstory canopy tree densities. It is only recently that gradient studies have become more common.

The assumptions of experiments using traditional agricultural experimental designs include high within-treatment unit homogeneity and provide a strong incentive for researchers to establish their studies on uniform or very comparable sites. Experiments with highly homogeneous conditions are statistically more powerful in finding significant treatment effects. Any review of the literature in silviculture in academic journals such as the *Canadian Journal of Forest Research*, *Forest Science*, or *Forest Ecology and Management* will show that silvicultural researchers aim to select sample plots that are as uniform as possible with respect to their soils, slope, aspect, and disturbance history for testing experimental treatments (e.g., q-factor, planting stock types, vegetation control levels, thinning densities). In our experience, finding uniform areas to test new silvicultural practices is often the most difficult task when implementing experiments, especially when working in unmanaged forests. For example, the optimal experimental setup to examine influences of stand density on tree growth would have perfectly uniform site conditions across all sample plots combined with minimal genetic variation among study trees. In practical terms this results in thinning studies being limited to the interior portions of single-species stands. Multispecies stands, stand edges, gaps, disturbed areas, or unique areas such as wetlands and riparian zones are carefully avoided to decrease variability within the study, even though they may be a vital part of the landscape.

Just as within-treatment variability in site condition and study objects is undesirable in experiments, the statistical approach also requires researchers to rigorously control any external factors that might influence experimental treatments. For example, in a long-term spacing trial designed to determine optimum planting densities to maximize merchantable volume, researchers might build a fence to protect seedlings from browsing damage. Similarly, any trees affected by insects or disease would be excluded from the analysis. Studies in which variation due to other exogenous (nontreatment) factors is very large are considered problematic because they interfere with the ability to accept or reject the null hypothesis. Frequentist statistics thus encourage researchers to minimize the variation of all factors with the exception of the experimental treatment.

The characteristics of agricultural experiments discussed above further encourage homogeneity in management as they promote studies with a limited inference scope. Information about the range of conditions (e.g., site type, aspect, elevation, species) to which study results apply is the scope of inference of an experiment. If the inference scope is narrow, results should be applicable only to those narrow conditions. Designed agricultural experiments have to consider the balance between statistical power to find difference and wider applicability of study results (Ganio and Puettmann 2008). Typically, researchers first decide on their inference scope and then lay out an experiment to ensure that treatment conditions across replications reflect the inference scope. The frequentist statistical approach is more likely to find treatment differences when the variation in external factors and the resulting experimental error are small. This will be the case when replicates are more similar; that is, the inference scope is small. For example, vegetation control studies are more likely to find significant impacts of competing vegetation when the study sites all have the same moisture and nutrient conditions.

Intensive highly controlled silvicultural studies can likely cover only a small portion of sites and will not necessarily reflect all the variability in conditions found in natural forests or even most managed stands. Natural forests and plantations are almost always much more heterogeneous than the experimental conditions where a particular treatment is tested. Most silvicultural publications do not provide specific descriptions of the inference scope (but see Cissel et al. 2006). Instead, information about the inference scope must be gleaned from study site descriptions. It is typically left up to readers of scientific reports to decide whether the study conditions are similar enough to their area of interest to make the study results applicable. Consequently, practicing silviculturists had to become comfortable with applying best treatments based on information from a limited number of experimental studies, often with very small inference scopes.

The use of traditional agricultural experimental designs and the search for best treatments has had a profound but largely unrecognized influence on how forests are managed throughout the world. Probably the greatest influence of the agricultural research model on silvicultural thinking was the implicit message that an identifiable best treatment or suite of practices exists for a particular management situation. When silviculturists attempt to reproduce results achieved in experimental studies on larger scales, such as landscapes, the agricultural research model encourages them to apply the best treatments consistently to all stands, rather than to embrace or adopt a variety of different silvicultural approaches. The adoption and dominance of the agricultural research model has not led to a culture of trial, innovation, or examination of trade-offs among practicing silviculturists, but has supported a conservative culture of implementing standardized prescriptions.

The history of implementation of silvicultural systems around the world provides an appreciation of the influence of the agricultural research model on contemporary silviculture. First, many aspects underlying the agricultural research model were already well-established in forestry long before the development of scientific silvicultural research. For example, silvicultural systems were descriptive management systems that included the harvesting, regeneration, and tending methods needed to create specific types of even- or uneven-aged stands. They already had many characteristics that later became indicative of the agricultural research model, including a limited set of treatments, a bias toward uniformity, and a focus on mean responses.

An important distinction needs to be highlighted. In Europe, where the individual silvicultural systems evolved well before the development of the agricultural research model, application procedures did not focus on widespread applications of a single best treatment (with notable exceptions, see chap. 1). European silviculturists are still more apt to incorporate small-scale variability into individual systems based on long-term observations, local experience, and new ecological knowledge (e.g., Pommerening and Murphy 2004). In contrast, application of silvicultural systems outside Europe, for example in Canada or the United States, began mainly after the agricultural research model had become solidly entrenched in silvicultural thinking. Individual silvicultural systems were thought of in terms of a prescribed program of fixed treatments and, in general, local modifications and adjustments were not encouraged. Furthermore, throughout the twentieth century, educational material relied on scientific studies that determined best treatments for particular species or regions. For example, the series of U.S. Forest Service manager's handbooks in north-central states (e.g., Benzie 1977; Perala 1977; Sander 1977) provided silviculturists with fixed sets of possible treatments for the major commercial tree species. These guides and other subsequent guides were powerful teaching tools and provided students and practicing silviculturists with a quick way to become familiar with local silvicultural constraints and opportunities without necessarily having to visit the woods. On the other hand, such guides further ingrain the belief in a best treatment; they emphasize knowledge over thinking, and are not designed to encourage innovation or local adaptation as an inherent part of practice.

In many parts of the world, the widespread application of uniform silvicultural systems combined with the use of designed experiments to identify best treatments for individual practices or suites of practices has resulted in fairly homogeneous conditions in terms of tree species and stand structures within and among managed stands. This is especially the case for plantation management, for example large-scale industrial forestry operations in North and South America, where the same species is planted at the same density on tens to hundreds of thousands of hectares. But, it also applies to other even-aged systems and uneven-aged forest management systems where variability is purposefully reduced and controlled through management. Even the *Dauerwald* movement (see chap. 1) or its derivations, close-to-nature forestry, minimizes variation within and among stands by emphasizing a limited set of possible stand structures for all stands and conditions.

The Scale-Independent View of Forestry Practices

The assumption of scalability is implicit in agricultural experimental designs and has also influenced how silviculture relates to homogeneity. Much of the silvicultural science and management has been within the disciplinary structure of universities and government forest agencies responsible for forest management. Within this disciplinary structure, there are established, though constantly evolving, norms for good science and management. As previously discussed, silvicultural science has been heavily influenced by the agricultural research model resulting in the strong belief that information describing structures, relationships, or processes in forest ecosystems can be derived from small experimental plots and then be easily scaled up to stand or landscape levels.

Researchers working in small and very homogeneous plots are not concerned about scaling up when experimental conditions are closely reflecting situations where the results will be applied. In these instances, calculating the average response on small plots likely provides information applicable to similar but much larger units, for example, agricultural fields. As silvicultural researchers adopted this research model, they implicitly accepted that the study of practices in small plots provides reliable information to guide management at much larger scales. This assumption of linear scaling further influences how silviculturists viewed homogeneity in forest ecosystems. If the assumption of uniformity across scales is met, results from small research plots can be scaled up and operational practice would be expected to yield the same results as the designed experiment. Being able to use scientific findings only by scaling up sends the message that study conditions (i.e., uniform stands) are the norm and an inherent requirement of good "scientific" forest management.

With very few and mostly recent exceptions (e.g., see listing of large-scale experiments in chap. 4), silvicultural research plots were much smaller and more uniform than the stands to which the results were expected to be applied. Most silvicultural studies during the 1960s to 1980s utilized small plots (e.g., 0.1-hectare plots for the Level of Growing Stock Study, Marshall and Curtis 2002). From an experimental viewpoint, the use of small plot sizes had several advantages. It made it easier to locate homogenous areas and to increase the number of replicates. It allowed more efficient use of land, labor, and other resources needed for research. Scalability from research plots to managed stands was further enhanced by use of scale-independent measurement units (e.g., trees per hectare) that could be directly translated into stand-scale activities (see earlier discussion on stand management and stand descriptors).

Discrepancies between results of applications in small, highly controlled growth and yield research plots versus stand-scale applications have been known for a long time. For example, Bruce (1977) suggested that a solution to the problem is to make managed stands more uniform and thus more similar to the research plots. In effect, the problem of scaling up encouraged and promoted the management of homogeneous stand conditions.

Large-scale operational application of new silvicultural treatments that proved superior under limited study conditions can also produce different results than predicted. For example, the yield that can be expected from managing the sugar maple (*Acer saccharinum*) forests of Québec by the single-tree selection system (*coupes de jardinage*) has been carefully studied using replicated experiments (Bédard and Brassard 2002). Operational implementation of the treatment that proved best in the experiment did not produce the predicted results when applied widely by forest companies. Physical damage during logging, thinning shock, and individual stem mortality due to windthrow were found to be, on average, much higher in operational areas than in the experimental setting. Some operational stands produced similar results to those found experimentally, but overall there was considerably more variability in the operational logging, resulting in greater variability in yield. Scaling average responses from small experimental plots can be inadequate to characterize and understand important processes that control growth responses in naturally diverse forests. A general analysis of scaling-up issues continues to receive little attention in silviculture research.

An alternative approach to research that averages variability and focuses on uniform application at the stand scale is to tailor research and prediction to the spatial scale of interest. For example, one of the most important events silviculturists must understand and predict is the recruitment of new tree seedlings, which likely needs to be studied at multiple spatial scales that are not necessarily related (Houle 1998). Seed availability is largely influenced by the nearby abundance of parent trees acting on spatial scales ranging from a few meters for heavy-seeded species (e.g., oaks, chestnuts) to a few hundred meters for the vast majority of species with lighter, wind-distributed seeds (Greene et al. 2004). Seed dispersal distances, and therefore the appropriate scale of study, can be further influenced by stand structure (Clark et al. 1998; LePage et al. 2000). Alternatively, seedbed substrate varies at the microsite scale, but substrate favorability can also be strongly influenced by local canopy structure (Cornett et al. 1998; LePage et al. 2000).

The study of tree growth in small uniform plots can lead to the conclusion that competitive forces are applied equally throughout the stand, which encourages the viewpoint that spatial variability at scales smaller than stand-level is not important. For example, growth and yield researchers have repeatedly tested whether integrating small-scale spatial variability in growth models improves model predictions. In comparative studies of distance-independent and distance-dependent competition indices, they generally concluded that spatially explicit, distance-dependent competition indexes provide no worthwhile improvement over spatially independent models (Daniels 1976; Alemdag 1978; Lorimer 1983; Martin and Ek 1984; Daniels et al. 1986; Corona and Ferrara 1989; Holmes and Reed 1991; Wimberly and Bare 1996). Results of comparative studies suggest that the spatial configuration of trees within a stand is not important for predicting individual tree and stand-level growth. Among other possibilities, this conclusion likely highlights the limited spatial and size variability found in plots utilized in these comparative growth studies. As discussed earlier, when studies use the agricultural model to investigate impacts of stand density on growth and yield, it is desirable to keep other factors, such as spatial arrangement, as homogeneous as possible. Thus, unless specifically designed to investigate spatial arrangement, the research approach is biased against accounting for the effects of within-stand spatial variability. The generally accepted validity of many growth models that assume de facto regular spacing leads to the impression that small-scale spatial variability is not important in influencing stand development and has resulted in the belief that fine-scale spatial variability can be ignored when managing forest stands.

Competition Indexes: Most growth models do not explicitly account for the presence of spatial structure in tree data, but rather use competition indexes to incorporate information about a subject tree and its neighbors. Distance-independent indexes are simply functions of stand-level variables or dimensions of the subject tree. Distance-dependent indexes use neighborhood-scale information in an attempt to capture fine-scale changes in competition due to the distance between the neighbors and the subject tree and their relative or absolute dimensions. See Moeur 1993; Shi and Zhang 2003; Stadt et al. 2007.

Focus on Predictability

In general, over the last two centuries silviculturists have successfully provided a steady and predictable flow of timber and wealth. To accomplish this, silviculturists had to limit the influence of stochastic disturbances, refine regeneration and stand tending practices, and emphasize homogeneous stand conditions. These practices also reduced the variation in stand-level responses. One key reason for homogenizing the temporal, spatial, and structural components typically found in natural forests was the need for increased predictability of stand development and therefore of yield.

Efforts to predict yield have always been crucial for assessment of silvicultural practices. Since its very beginnings, the historical development of silviculture has been closely linked with concerns to ensure sustainability of wood supply (von Carlowitz 1713) and these needs led to the development of the normal forest concept (see earlier discussion and chap. 1) and other tools for forest planning. For example, in the early twentieth century in parts of North America, "Hanzlik's formula" was applied to ensure that ongoing harvest rates resulted in the conversion of forest estates to a normal forest and that equal annual volumes of timber were available in perpetuity (Hanzlik 1922). By now, most regions have moved beyond Hanzlik's formula to include social, economic, and environmental considerations in their calculation of wood supply.

The calculation of a sustainable harvest rate requires reliable information about tree and stand growth through repeated inventories, growth and yield models, or some combination of the two. It also requires silvicultural practices that ensure reliability and consistency of regeneration and tree and stand growth patterns. To ensure timely natural regeneration, early silviculturists developed reproduction methods to promote and enhance a reliable seed supply and to provide optimal conditions for the natural establishment of preferred tree species (e.g., seedtree or shelterwood; Matthews 1989; chap. 1).

Developments in the United States and Canada during the late twentieth century provide good examples of how large-scale industrial logging activities impacted the reliability of natural regeneration and how, in turn, these concerns were addressed by silviculturists to ensure predictable regeneration (see Cleary et al. 1978; Lavender et al. 1990; Wagner and Colombo 2001). In many parts of North America, natural regeneration was considered not consistent enough. To improve reliability and predictability of regeneration in regions where clearcutting large areas became a widespread practice (e.g., Weetman and Vyse 1990), silviculturists developed tree nurseries and planting programs for selected tree species and increased research efforts to ensure more consistent reforestation than naturally occurs after harvesting (Thompson and Pitt 2003).

As part of these efforts, the regeneration phase, from seed storage to germination and early seedling growth, was moved into tree nurseries or greenhouses. Rather than allowing for stochastic elements such as predation or weather to influence early seedling establishment, these factors were controlled. Greenhouses provided a perfect, climate-controlled setting where light, nutrient, and water levels could be managed. With proper seed collection and storage, germination conditions, and protection from insects, diseases, and weeds, nurseries became efficient at producing reliable and homogenous planting stock. Planting tools, site preparation techniques, and vegetation control practices were refined to ensure a high survival rate of planted seedlings. In regions with intensive forest management, the combination of vigorous planting stock, site preparation, and vegetation control regularly results in higher than 90 percent survival of planted seedlings.

Efforts to improve predictability also focused strongly on aspects of tree and stand growth (Rudolf 1985; Curtis et al. 2007). Inventory plots and growth and yield experiments were installed in response to the need for long-term predictability of tree and stand growth. The development of growth models followed in some regions. To promote predictability and reflect "ideal" management scenarios, growth models were mostly based on data from small, uniformly structured research and inventory plots (see also scale discussion earlier). Furthermore, when data were used in the analysis of studies or pooled from various studies to develop a regional growth model, only those sample plots that had maintained their integrity (had experienced no or limited disturbances) were used in the analysis (e.g., Buckman 1962; Pretzsch 2005). It was not uncommon for individual trees, plots, or entire replicate units to be dropped from growth and yield experiments if outside factors such as herbivory, disease, or windstorms increased variation, thus reinforcing the notion that managed forests should be free of unplanned disturbances. In reality, it may be rare for any stand, managed or unmanaged, to remain totally free of insects, disease, or storm damage for extended periods of time.

Most early yield tables and growth models were capable of making predictions only for single-species even-aged stands due to a combination of the use of agronomic study methods and limited computer power. Many models used today to predict growth rates still have that limitation, which creates an interesting dilemma. If determining sustainable harvest levels is deemed important, and reliable growth predictions are available only for single-species simple structured stands, then simple structured stands are favored by silviculturists. This dilemma can be avoided by investment in a sophisticated permanent inventory of a wider range of stand types (e.g., continuous cover forestry) or development of more complex growth models. In general, the restriction of growth models to predicting yield under only uniform conditions has encouraged the simplification of practices and homogenizing of structures in managed stands. The measurement of growth in permanent inventory plots may not have the same limitations as single-species growth models in terms of dealing with mixed-species stands. However, as long as inventory plots aim to determine maximum sustainable yield levels, the underlying premise still reflects silvicultural thinking that fully stocked, evenly spaced stands are the norm or reference condition and deviation from this norm is then considered bad forestry.

This norm or reference condition on which yields are projected may be an artificial or idealized condition that doesn't actually exist. For example, almost half (45 percent) of wood harvested in 2004 on intensively managed state land in Baden-Württemberg, Germany, was unplanned and in response to disturbances (Anonymous 2005). Indirect effects of climate change, such as when responses of one species to a climate trend in turn affect different species, provide another example. Woods et al. (2005) describe strong evidence that the fungus *Dothistroma septosporum*, in response to a directional increase in summer precipitation, is negatively impacting lodgepole pine plantations in a completely unexpected way. Yield projections need to be reassessed for extensive well-stocked pine plantations, now defoliated or dying because of the fungus after an increase in summer precipitation, which would be expected to favor tree growth.

Small- and larger-scale disturbances are an integral part of a landscape, and their effects on stand development cannot be predicted from growth models that assume fully stocked, regular stands. Most forests exhibit a pattern of disturbance-induced change that spans virtually all scales of space and time (Frelich 2002; Kimmins 2004; Johnson and Miyanishi 2007). If the norm is a fully stocked, homogeneous stand, disturbances are necessarily viewed as an external factor that negatively influences stand development, rather than as an integral part of stand and landscape development. This also creates an interesting discrepancy between the effort put into producing growth models with high accuracy and the rough corrections that are often used to account for the impact of stochastic elements.

The emphasis on predictability could be addressed by silviculturists only through control and homogenization of forest structures, and this focus has infiltrated all aspects of silviculture. The resulting top-down, command-and-control approach to silviculture is still deep-rooted in the discipline and difficult to overcome. The focus on predictability is not unique to silviculture and forestry. It is observed in most, if not all, renewable resource management disciplines that involve a harvest of a surplus (e.g., yearly harvest levels for wildlife and fisheries management).

The histories of forestry, fisheries, and wildlife management share similar patterns in this regard (Ludwig et al. 1993; Hilborn et al. 1995; Bottom et al. 1996; Struhsaker 1998). However, the link between predictability and a top-down, command-and-control approach weakens as increasing computer power, computational skills, and new technologies allow development of more sophisticated growth models and inventories that do not rely on homogeneous stands or the normal forest concept.

Command and Control: The tendency to apply increasing levels of top-down management to natural resources. It manifests itself in attempts to control ecosystems; and when ecosystems act in ways that are considered erratic, even more control is applied. Command and control often, however, results eventually in unforeseen consequences for ecosystems. The pathology of natural resource management is the loss of ecosystem resilience when the range of natural variation in the system is reduced. If natural levels of variation in system behavior are reduced through command and control, the system becomes less resilient to external perturbations, resulting in crises and surprises (Holling and Meffe 1996; Folke et al. 2004; Drever et al. 2006).

Conclusion

Silvicultural practices over the past few centuries have been adapted to a wide variety of objectives and conditions, but throughout its development silviculture has relied on several core principles. First, it has been predominantly tree-focused in application and assessment of practices. Second, it treated stands as homogeneous entities. Third, it utilized the agricultural research model in evaluating old and new practices. Fourth, it assumed that spatial scales are unimportant and that stand-level assessment and management were appropriate for all situations. Finally, it focused on achieving orderly and predictable forest development. These principles cannot be viewed in isolation from each other and from the influence of long traditions in silviculture. In conjunction, they have directly and indirectly affected how research is undertaken and have profoundly influenced how silviculture is taught to students and how practicing professionals think and act.

The shortcomings of the reliance on the above-described principles have become apparent with increased interest in a wider variety of ecosystem values, processes, and functions and a better understanding of forest ecosystems, especially of ecosystem health, productivity, and resilience. The current approach to silviculture research and management as described in the five principles has inherent characteristics that promote uniformity and discourage variability. This, in turn, has resulted in many managed forests having uniform or narrow ranges of tree species composition and stand structures. Thus silviculture, with its desire to control nature and ensure predictability, is an example of a discipline that has slipped into what Holling and Meffe (1996) termed the "pathology of command-and-control management in the natural resources." Furthermore, the reliance on long traditions and the associated conservative culture of silviculture has made it especially hard for silviculturists to respond to rapidly changing ecological knowledge, management objectives, or social views of forests.

3

Ecology

Acknowledging Complexity

Ecology is a young but well-established discipline in the biological sciences. Ecologists in Europe and North America have a long history of organizing themselves in professional societies. The first meeting of the British Ecological Society was held in 1913 and the Ecological Society of America first met in 1915. Ecology evolved almost directly from the rejection of the traditional descriptive approach to scientific work in biogeography (Harper 1982). Unlike forestry and agriculture, ecology did not develop from the need to address a practical problem. It developed from the desire to understand how species are distributed in the world and how they coexist. Ecology describes patterns in nature and strives to identify the mechanisms underlying those patterns.

The objective of this chapter is to review the main developments of the science of ecology, especially those related to the importance of complexity to ecosystem functions and processes, such as adaptability to altered conditions, biodiversity, resiliency, and productivity (Gunderson and Holling 2002; Scherer-Lorenzen et al. 2005; Drever et al. 2006). We further focus on research that has formalized concepts that have resulted in a better understanding of how ecosystems, especially forests, self-organize to produce complex patterns (*sensu* Levin 2005; Solé and Bascompte

2006). The notion of complexity has always been omnipresent in ecology and has influenced greatly the theory and tools used to study the natural world (Bradbury et al. 1996). We explore how concepts related to ecological complexity and the broader science around complexity theory are of value to silviculture in chapter 5.

Origin of Ecology

The science of ecology could not have developed without a basic understanding of evolution and what Darwin called the "struggle for existence." One of the main goals of ecology is to understand this "struggle" and how it allows so much life and resultant complexity in ecosystems to exist. Ecology has even been labeled the science of complexity, and the changing understanding of complexity has influenced methods of ecological research (Bradbury et al. 1996). In contrast to evolutionary science, which aims at understanding how this struggle has given rise to so many different species over time, ecology tries to understand the current diversity of life forms and the processes that allow for such ecological complexity. The word "ecology" was first mentioned by a German zoologist, Ernst Häckel (1834–1919). When reading Darwin's writings, Häckel decided that a new term was needed for the study of the extraordinary complexity of life forms and functions on the planet Earth.

In North America, the botanist Frederic Clements (1874–1945) is often viewed, especially by plant ecologists, as one of the leading individuals who helped to define this new science. Clements and several of his contemporaries, including Eugenius Warming (1841–1924) and Henry Chandler Cowles (1869–1939), were the first to combine theoretical principles with quantitative methods (albeit rudimentary by today's standards) to address questions regarding relationships between organisms and their environment. One of the first North American ecology textbooks focused on the challenge to develop quantitative approaches to understanding nature and was titled *Research Methods in Ecology* (Clements 1905). Ecology has developed from its beginnings as a science that utilizes experimental and mathematical methods to investigate relationships between organisms and their environments, community structure and succession, and population dynamics (Kingsland 1991), and ecologists continue to debate the best types of statistical analysis to represent natural systems (Hobbs et al. 2006).

Review of Past and Current Concepts in Ecology

A brief historical review of ecological concepts related to complexity highlights a progression toward a more detailed and sophisticated understanding of the notion of complexity in ecological systems. Table 3.1 provides a timeline of selected ecological concepts which, while not all-inclusive (Keddy 2005), covers the milestones in the development of ecology. We describe these concepts as well as selected philosophical and technological advances in related sciences that have influenced the development of ecology.

Succession was one of the most influential concepts developed by ecologists in the late nineteenth century and became formalized under the leadership of Clements (1936). In response to questions about patterns of vegetation change over time, it stated that, after disturbances, plant communities (the ecosystem concept was not yet popular) develop toward a stable equilibrium called climax (Clements 1936). Clements's view was that this stable climax state represented the final and highly evolved stage of plant communities, and thus was more desirable than younger successional stages. This point of view, which underlies contemporary perspectives on old-growth forest conservation, resonated not

Ecological Succession: Refers to (to some extent) predictable and orderly changes in the composition and structure of plant communities in the absence of disturbances. Primary succession occurs in a new or unoccupied habitat, and secondary succession is initiated following disturbances that leave some of the biota intact. Succession was formerly seen as reaching a stable end-stage called the climax. It is now recognized that all ecosystems are in a non-equilibrium condition, although some stages—often the later stages—change more slowly than others.

only with ecologists but also with society at large, presumably because of its strong parallels to widely held aspirations for human progress and civilization.

Table 3.1. Chronology of important ecological concepts

Ecology Concept and Tools	*Main Proponent*	*Date*	*Forestry Concepts*
Competition	Darwin	1860	Vegetation management
Niche	Elton	1925	Regeneration
Succession	Clements	1930	Stand development
Ecosystem	Tansley	1935	Ecosystem management
Competitive exclusion	Gause	1935	Self-thinning
Environmental gradient	Gleason	1940	Plantation
Food-web and energy cycling	Odum	1955	Forest productivity
Multivariate analyses	Bray and Curtis	1960	Forest classification
Plant population/plant plasticity	Harper	1965	Genetic improvement
Coevolution/maximize fitness	Ehrlich and Raven	1965	Pest management
Island biogeography	MacArthur	1965	Landscape planning
Diversity-stability	Margalef	1970	Emulating natural disturbance
Resilience	Holling	1975	Emulating natural disturbances/landscape planning
Chaos	May	1970	—
Community computer models	Botkin	1975	Growth and yield/stand dynamic
Biodiversity	Wilson	1980	Variable retention/ecosystem management
Gaia	Lovelock	1980	Emulating natural disturbance
Meta-population	Hanski	1985	Landscape planning
Disturbance	Pickett and White	1985	Emulating natural disturbance
Scaling issues	Holling	1990	Landscape planning
Neutral theory	Hubbell and Bell	2000	—
Complexity	—	2000	—

Clements's view of succession cannot be appreciated without an understanding of competition theory, which is perhaps the ecological concept with the most influence on silviculture. Early in the twentieth century, competition was perceived as decreasing during the successional development toward the climax community. Early competition theories suggested that populations of plants competed more intensely during the early stages of succession and over time the allocation of resources among species and individuals became more efficient, resulting in less competition. This view explained observations that similar plant communities developed after longer periods with disturbances in different regions. Competition theories suggested that certain groups or associations of species were more compatible with one another and since competition was lower in later successional communities, they were able to develop harmonious and enduring relationships (Clements 1936).

Clements's view of succession also implied that disturbances were an external and undesirable phenomenon that interfered with the progress of plant communities toward the desirable or stable climax state, a view that was entrenched in the discipline until the 1980s (Pickett and White 1985). While much discussion occurred about whether Clements viewed climax communities as a quasi super-organism comprised of many interrelated parts, each vital to the functioning of the entire community, most ecologists agree that he understood late successional communities as having developed greater interdependence than early successional communities.

This view was first challenged by Tansley (1871–1955), who invented a new term for this complex state or system in which various organisms are—at least partially—interdependent with each other, the *ecosystem* (Tansley 1935). Gleason (1882–1975) carried the criticism further by suggesting that plants did not associate to form stable and predictable communities or ecosystems, but rather that they individualistically occupied positions along *environmental gradients*. By suggesting that each plant species possesses specific ecological requirements and that those species with overlapping requirements can be found growing together as a more-or-less random outcome of their individualistic habitat preferences, Gleason laid the foundations for the niche concept and foreshadowed the development of the neutral theory.

Consequently, Gleason disagreed with the concept of plant associations. Instead, he viewed plant communities as changing continuously over topographic and climatic gradients (Gleason 1926). Gleason suggested that plant assemblages were present not because plant species formed an interdependent entity, but because they shared similar biophysical requirements. Barbour (1996) proposed that Gleason's view may have replaced Clements's because it fit better with the prevailing North American geopolitical values of free enterprise, where each person is solely responsible for their actions and positions in society. Clements's view may have been considered too close to communist principles in which the government exerts a tight control on society and individualism is stifled and discouraged. Such cultural and philosophical influence was less prominent in Europe, and Clements's view influenced the Braun-Blanquet approach of plant classification (Braun-Blanquet 1928). This classification system developed a hierarchical taxonomy for plant communities along the same lines as the Linnaean taxonomy used for organisms.

At the same time in North America, Whittaker (1920–1980) endorsed the concept of ecological gradients. He subsequently developed a series of analytical techniques that allowed ecologists to study natural ecosystems along gradients to complement analysis techniques that relied on the view of ecosystems as discrete communities (Whittaker 1956, 1967). These new techniques became the precursors of *multivariate analyses* as developed by a variety of plant ecologists in different countries (e.g., Sorensen 1948, Denmark; Goodall 1954, Australia; Bray and Curtis 1957, United States). The development and prominence of multivariate analysis techniques in ecological research crystallizes the perception of nature as complex and driven by multiple variables.

While plant ecologists were refining their analytical techniques, animal ecologists were proposing the *niche concept* (Elton 1927; Hutchinson 1957) as a tool to understand the structure and functioning of ecosystems. The niche concept suggests that each species occupies a zone or habitat within which it can outcompete other species (see Silvertown 2004 and Chase and Leibold 2003 for recent reviews). The diversity of species could be explained by the niche concept and the fact that each species was better adapted to a certain portion of the ecosystem than others. Directly related is the concept of *competitive exclusion* (Gause 1934), whereby Gause developed mathematical techniques to describe how species that compete for the same resource cannot coexist. The strong mathematical influence in ecology was reinforced by Lotka and Volterra (Begon et al. 2006), whose equations are a standard topic in many ecology classes. The competitive exclusion principle provided

---

Ecological Niche: Describes the range of habitat conditions that a species or population can occupy within an ecosystem. The niche represents a multidimensional temporal and spatial space, where the biological, physical, and chemical environment is suited for a species. The fundamental niche of a species refers to the range of habitats it can potentially occupy in the absence of interference from other species. The realized niche, which is necessarily narrower, describes the actual range of habitats occupied by the species in the presence of competitors.

---

ecologists with a mechanism to explain how communities structure themselves. It also allowed a better understanding of the process of succession as species replacing one another as the environment is progressively modified by each set of species.

The central role of niche theory in ecology has been challenged recently by a new concept, the *unified neutral theory*. It states that species coexistence and patterns of abundance and distribution within ecosystems are governed more by the stochastic processes of extinction, immi-

---

Unified Neutral Theory: States that autoecological differences between members of an ecological community (e.g., birds, trees, moths, and so on) are much less important for a species's success than suggested by the niche theory. Instead, it stresses the rate of immigration and speciation and local stochastic (essentially random) processes that cause mortality and regeneration. This theory can be viewed as a null hypothesis for testing the niche theory and claims to better predict the diversity and relative abundance of species in various ecosystems.

---

gration, and speciation than by intrinsic ecological differences of species (Hubbell 1997, 2001; Bell 2000; Fargione et al. 2003; Volkov et al. 2003; Chave 2004). Hubbell's theory has stimulated new rounds of experimentation to quantify the importance of niche differences (e.g., Gilbert and Lechowicz 2004; Gravel et al. 2006).

Among plant ecologists in the second half of the twentieth century, John L. Harper stands out for his role in developing the subdiscipline of plant population ecology and the use of controlled experimentation to understand how plant communities function (Harper 1967, 1977). He developed concepts such as *plant plasticity* and the relative importance of vegetative versus sexual reproduction. Harper's work exemplified major advances in understanding how individual species evolved plasticity in relation to their environments. At the same time, an appreciation was also emerging that species did not evolve independently. The concept of *co-evolution* describes evolutionary developments when species evolve together over long periods of time (Ehrlich and Raven 1964). Although the notion of coevolution did not rehabilitate Clements's view of ecosystems as superorganisms, it suggested that species that evolve together for a long time may develop strong interdependencies that represent more than the simple addition of individual species characteristics. An important concept related to evolution is the development of optimal traits that *maximize fitness*. It is a long-held belief by ecologists that evolutionary selection pressures favor species that optimize their chance of survival.

Coevolution: Mutual evolutionary influences (either negative or positive) that exist between species and exert selection pressure on one another. Over time, each species evolves in direct response to the influence of other species. In some cases, a very strong mutualistic (e.g., mycorrhizae; mechanisms to ensure pollenization) or antagonistic (e.g., tolerance by herbivores to toxins produced by their host plants) relationship can develop.

As ecologists gained a better appreciation about the diversity of species, ecosystem processes, and the mechanisms controlling this diversity, they became interested in the function of diversity in ecosystems. Questions such as whether species are redundant and whether diversity makes ecosystems more resilient to change became of interest, leading to the investigation of the *diversity-stability* relationship (Margalef 1969). At this time, disturbances were still mostly perceived as setting back or hindering ecosystem development (a carryover of Clements's ideas) and research addressed the question of whether diversity of species and ecosystem processes and functions can act as insurance against disturbances and environmental fluctuations. This area of research has since been a topic of intense discussion (McCann 2000). For example, using chaos theory, May (1975) showed mathematically that diversity does not guarantee stability. More recently, research by Tilman and others (Tilman et al. 1997; Tilman 1999, 2004) has shown that species diversity is important not only for stability, but also for productivity of ecosystems. Discussions about the validity of diversity-stability relationships are still ongoing and partially fed by different definitions of concepts such as stability, complexity, and diversity of systems. A recent book by Loreau et al. (2002) concluded that strong scientific evidence exists that biodiversity enhances functioning and stability of ecosystems (fig. 3.1), leading to concerns that any loss of diversity could negatively affect long-term ecosystem functioning. This issue continues to be a topic of debate and ongoing research in contemporary ecology (Hooper et al. 2005).

Ecologists had to develop new approaches for ecological study in response to the concept of ecosystems being accepted (Odum 1969). Studies had to be large enough to encompass an entire ecosystem in order to investigate questions such as how manipulations of ecosystems influence basic ecological processes. The Hubbard Brook experiment (Likens et al. 1970) and the International Biological Program (IBP) are the most renowned examples of efforts in forest ecosystems to test ecosystem-level hypotheses. Ecologists developed novel approaches in their efforts to understand what factors influence the development of diversity. A major breakthrough was the experimental study of isolated ecosystems such as islands, which showed that larger and less isolated islands had higher species diversity. This resulted in the *island biogeography theory* (MacArthur and Wilson 1967; Simberloff and Wilson 1969) that species diversity in isolated patches, such as islands, is governed by immigration and extinction rates. The field of conservation biology, with its emphasis on topics such as landscape connectivity, fragmentation, and reserve design (Simberloff 1988; Hunter 1990; Seymour and Hunter 1999), was built on a theoretical foundation of concepts such as island biogeography, issues of scale (Levin 2000), and *metapopulations* (Hanski 1999). To better address these concepts, ecologists developed terms for different aspects of diversity, including alpha (or within-habitat diversity), beta (or between-habitat diversity), and gamma (geographic diversity). As the assumption of disturbances as outside agents of ecosystem development was questioned,

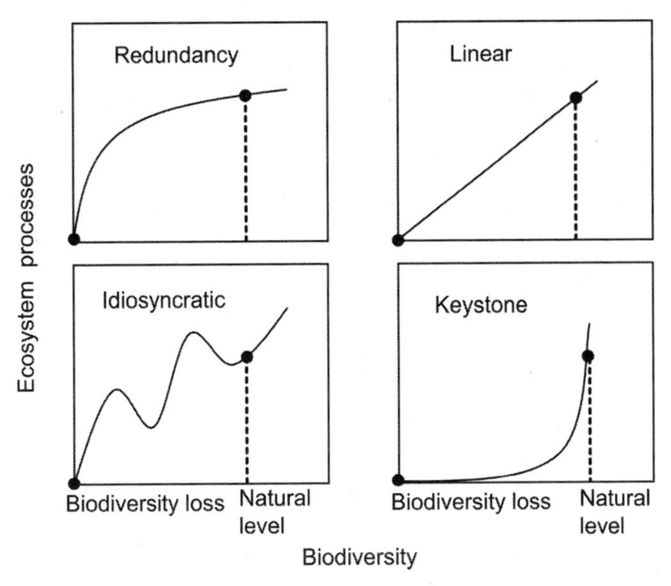

Figure 3.1. Graphical representation of hypothetical relationships between biodiversity and ecosystem processes. The "natural level" of biodiversity (dashed line) represents typical or mean values observed in unmanaged reference ecosystems. Below that level, the system experiences biodiversity loss. The redundancy relationship shows a system that reaches its normal functionality at a low level of biodiversity. The linear relationship shows a system that continuously increases its functionality with increasing biodiversity. The idiosyncratic relationship shows a system that changes its functionalities in an irregular manner. Finally, the keystone relationship shows a system that depends on a special species to achieve functionality. (Adapted from Loreau et al. 2002.)

topics like the influence of size and frequency of disturbances on diversity became important (Frelich 2002). For example, the *intermediate disturbance hypothesis* (Connell 1978, 1980) suggested that the greatest diversity was found in regions where the disturbance level was intermediate.

Metapopulation: Consists of a group of spatially and physically separated populations of the same species that interact to a limited degree. It is related to the concept of island biogeography. The theory emphasizes the importance of connectivity between seemingly isolated populations and is used in conservation biology for planning networks of reserves.

Although it is ubiquitous today, the term *biodiversity* was coined by Wilson (1988) in order to integrate the various dimensions of species, structural processes, and functional diversity. Biological diversity is the result of thousands and even millions of years of evolutionary processes. One of the most important elements of biological diversity, the number of species, has varied a lot over the last 4 billion years, but the current rate of disappearing species is believed to be unparalleled. It is now recognized that maintaining biological diversity is an important element for the normal functioning of any ecosystem (Loreau et al. 2002).

Biodiversity: Although many different definitions exist, a straightforward one is "the variation of life at all levels of biological and ecological organization." It includes all genes, species, ecological and biological processes, and ecosystems of a region. Indexes traditionally used to described the biodiversity of a region include: (1) alpha diversity, which refers to the diversity of taxa within a particular area, community, or ecosystem; (2) beta diversity, which compares the number of taxa among areas, communities, or ecosystems; and (3) gamma diversity, which refers to the overall diversity of areas, communities, or ecosystems within a certain region of the globe.

As the science of ecology moved forward, ecologists recognized that just as single species evolve, communities and ecosystems are also evolving (or changing) in response to changes in the environment. Thus, the very notion that communities or ecosystems are defined as stable has been rejected. Instead, ecologists now perceive the ability to evolve or change in response to changes in environment or disturbances as a very important component of ecosystems. Without the ability to respond to change, most ecosystems would be expected to fail or vanish. To stay functionally "fit," ecosystems have to continuously reinvent themselves. In this context, *disturbance* cannot be perceived as an undesirable externality. Instead, disturbances are an integrated and important component of the ecosystem dynamics (Pickett and White 1985; Frelich 2002).

The shift in thinking is especially apparent in ecologists' views of fires. For a long time, fires were viewed as catastrophes, and land management organizations went through tremendous efforts in fire prevention. Now, fires are perceived by ecologists as an essential element that contributes to the health and functioning of ecosystems. As Connell (1978) describes, disturbances may influence community diversity by providing the necessary changes in conditions to maintain the diversity of species found in one region.

The focus on the development of ecosystems also put the importance of competition versus facilitation in a different perspective (Brooker 2006). In this context, ecosystems are similar to the way we

---

Gaia Theory: Proposes that all living and nonliving parts of the earth are functioning in a complex, interacting way comparable to organs in an organism. All living and nonliving parts of the earth are interacting to improve or at least maintain the living conditions on the planet. The theory suggests that life itself and many geophysical processes on earth have evolved in an orderly way to improve the livability of the planet. The idea is still debated today, but no mechanism is known that could explain such orderly regulation of life and processes at the scale of a planet.

---

perceive human society. Neighbors may be in competition for jobs and at the same time collaborate to make their neighborhood livable. An extreme view of these principles is Lovelock's (1979) *Gaia theory*, which suggests that all living organisms work somewhat together to regulate the environment of our planet.

Ecological Complexity and Complexity Science

The concepts and theories developed by ecologists to understand the origin and importance of biodiversity have led to the view that ecosystems such as forests are *complex systems* that fit within the definition of complexity theory (Gallagher and Appenzeller 1999; Parrott 2002). Accepting ecosystems as complex systems required a profound change in thinking by ecologists and resulted not only in new insights, but also in new challenges (Naeem 2002; Parrott 2002; Levin 2005; Solé and Bascompte 2006). Complexity emerges in ecological systems due to adaptation of and coevolution between organisms and their environments across multiple scales of space and time (Levin 2005). This suggests that the basic forces that control the evolution and adaptation of species constitute the basic mechanisms that create complexity in ecological systems. In contrast, heterogeneity or variability refers to biotic or abiotic

Complexity Science: Investigates how relationships among individual parts or single processes can give rise to collective behaviors of the whole system that cannot be predicted by its parts. The science of complexity is different than, but complementary to, the study of biodiversity. Complex systems have several defining features: (1) nonlinear relationships and indeterminate, chaotic or quasi-chaotic behavior make predictions uncertain; (2) boundaries are difficult to determine and we are never certain what defines the system; (3) the system is open to outside influences and so is never totally at equilibrium; (4) relationships contain feedback loops that may cross scales or hierarchies of organization, making the system self-regulated or self-organized; (5) the system can exhibit behaviors that are emergent—that is, behaviors that cannot be predicted from the individual parts of the system or from understanding the individual components of lower levels of organization; and (6) the system "remembers" its previous states, as prior states partially influence present ones.

characteristics that change greatly either spatially or temporally within a system.

Even basic issues such as how to study ecosystems require rethinking when ecologists accept that ecosystems are characterized by strong (usually nonlinear) interactions among various components, with complex feedback loops and significant time and space lags, discontinuities, thresholds, and limits. Complex systems are not well understood using classical or Cartesian modes of thinking, such as reductionist or determinist science (Gershenson and Heylighen 2003). Instead, ecologists had to rely on a different set of concepts and models to decipher ecosystems' properties and functions.

The *ecological resilience* of ecosystems (Holling 1973) is such a concept that could not have been developed from reductionist designed experiments, but rather evolved from integration of theoretical concepts and predictive models. The ecological resilience of an ecosystem is defined as "the capacity of an ecosystem to tolerate disturbance without collapsing into a qualitatively different state that is controlled by a different set of processes" (www.resalliance.org/576.php). Ecological resilience emphasizes persistence, adaptiveness, variability, and unpredictability. It recognizes that ecosystems are in nonequilibrium and that changes in ecological processes at one scale can affect other processes at other scales in unpredictable ways. The concept of ecological resilience is highly compatible with complexity science.

Holling (1992) further developed this concept to incorporate the challenge of *scaling in ecology* and determined that the study of ecosystems requires approaches that are specifically designed for the temporal and spatial scale of the respective question (for a recent example of the impact of scale on study results, see Gunton and Kunin 2007). Holling recognizes six scales in forest systems: (1) the leaf level, (2) the crown level, (3) the gap or patch level, (4) the stand level, (5) the landscape level, and (6) the biome level. In addition, Holling suggests that ecologists can make major contributions by focusing on linkages among scales, especially since studying or modeling ecosystems may require representation of lower-scale processes (e.g., number 1 or 2, as listed above) into larger-scale representations of ecosystems (e.g., numbers 4 and 5, as listed above). For example, to predict forest succession may not require simulation of the intracellular processes of photosynthesis and respiration, but likely requires an understanding of what resources influence growth and mortality of tree species. Research efforts are under way to develop rules of scaling that can help us develop a better understanding of how various processes work in a system and how they are connected to lower and higher scale processes (Rietkerk et al. 2002).

Another major turning point in the understanding of complexity in ecology was initiated by Robert May (1974). His exploration of complex and dynamic behaviors implicit in simple growth models popularized the *chaos theory,* which also has greatly influenced the discussion of complexity (Langton 1990). The maturation of chaos theory into a dominant intellectual movement was in large part due to the emergence of a general mathematical theory of nonlinear dynamic systems that also embraced chaos. While ecologists had a long tradition of using mathematics as a powerful research tool, the science entered a new phase when concepts

---

Chaos Theory: Describes nonlinear dynamical systems that can exhibit—in some conditions—seemingly unpredictable behavior. A key factor responsible for this behavior is the sensitivity to outwardly insignificant differences in initial conditions. We now understand that chaotic systems are deterministic in the sense that they are influenced by attractors, which determine the general direction or envelope of conditions in which the system will develop, although it is impossible to pinpoint exact locations.

---

and ideas were not only expressed in mathematical equations, but used as a research tool to simulate and investigate complex ecosystem dynamics. None of these developments would have been possible without advances in technologies, especially the increase in computing power.

Simulation models are useful tools to integrate and study the inherent features of complex systems. The first *community-level model* (JABOWA) of forest dynamics was as much an attempt to find uses for computers (Botkin et al. 1972a) as it was an investigation of ecological questions (Botkin et al. 1972b). It is quite telling that the JABOWA simulator was introduced to the ecological literature under the title "Some Ecological Consequences of a Computer Model of Forest Growth" (Botkin et al. 1972b). The approach taken by Botkin and coworkers acknowledged that forest development is a local spatial process. Utilizing data taken in small inventory plots and modeling development at the scale of individual plots (i.e., gaps), JABOWA simulated ecosystem-level development patterns, specifically key elements of tree succession. This approach (i.e., bridging scales) to understand community-level processes has proven very successful and JABOWA has inspired development of numerous community models for various types of ecosystems (Messier et al. 2003).

The debate is ongoing whether ecologists can explain the complexity of natural systems with better equations and increased computing power or whether nature is too complex to be simplified into models and only experiments can provide insights into complex behavior. Clearly, each approach has its strengths and weaknesses and more and more ecologists would argue that complex questions should be addressed using a variety of approaches, including field and greenhouse experiments.

Conclusion

The primary role of ecology is to understand how nature has produced such a diversity of life forms and structures. Ecology has made great progress in understanding natural systems and the importance of complexity in ecosystem processes and functioning. Maintaining heterogeneity, biodiversity, and complexity in forests is important to maintain all of their processes and functions. The role of complexity in natural systems for providing many essential goods and services (Daily 1997) and the suite of factors that are involved in regulating these goods and services (Loreau et al. 2002) are now better understood. Ecology, however, has struggled to translate these advances and insights into guidelines for management of natural systems that will keep the system functionally fit (see Peters 1991; McPherson and DeStefano 2003).

Ecologists have made some efforts to bridge the gap between the "fuzzy" concepts of ecology and management applications (e.g., Bazzaz et al. 1998; Palmer et al. 2004). Landscape ecology is a good example of such successful efforts. Also, the ecological understanding of the role of natural disturbances in ecosystem functioning (Pickett and White 1985; Perera et al. 2004) has resulted in integration of these concepts into forestry writings and practices (Hansen et al. 1991; Franklin et al. 1997; Seymour and Hunter 1999; Bergeron et al. 1999a, 2002; Kuuluvainen 2002). However, integration of aspects related to functioning of complex systems into management practices is still in its infancy.

We have included this discussion of ecological theories and concepts as they relate to our understanding of forests as complex systems with the aim of providing a starting point for silviculturists on these topics. To improve and move silvicultural practices forward, it is important for silviculturists to understand the theories about basic processes governing ecological functioning in forested ecosystems. The effective management of forests depends on ecological knowledge. We are not disputing that silviculturists have always considered their prescriptions to be based on some ecological understanding of ecosystems; otherwise silviculturists would not have been so successful at achieving ownership objectives (see chap. 1). However, ecological concepts are often "filtered" or interpreted and modified by silviculturists to fit within the current silvicultural approach and philosophy (Benecke 1996; Kerr 1999). We suggest that much—maybe even the main important component—of the significance of these new concepts (e.g., the benefit of embracing complexity) is lost in this process. The next two chapters will discuss these issues in more detail.

4

Silviculture and Ecology

Contrasting Views

The disciplines of ecology and silviculture have their own niche in the gradient from basic science to applied science and management. They offer different but complementary perspectives on how to manage the natural world in a sustainable way. This chapter builds on the previous chapters. Chapters 1 and 2 explained how silviculture evolved and developed with a strong emphasis on practices that promoted survival and growth of desired tree species in uniformly managed stands. Chapter 3 explored the historical development and progress of ecology, which has focused on understanding the natural world through the study of the interplay among species and processes at diverse spatial and temporal scales. In contrast to silviculture, ecology has not been directly concerned with the management of ecosystems (Bazzaz et al. 1998). However, theories and concepts developed by ecologists have indirectly influenced applications in silviculture (tab. 3.1). In this chapter we compare and contrast the views of the two disciplines and their interactions to offer insight into their respective strengths and limitations for developing new approaches to solving natural resource issues.

The disparity between the objectives of silviculture and ecology was apparently sufficient to allow establishment of the discipline of forest ecology. Forest ecology is a relatively young discipline that had its origin in forestry schools and was first taught in European and North American forestry programs "because of its importance in influencing silvicultural practices" (Spurr 1964, 7). Many topics presented in the first forest ecology textbook (Spurr 1964) and now considered the essence of forest ecology (except for tree genetics and physiology) were previously taught in forestry classes under the term *silvics*. Before Spurr's book, courses covering forest ecology were commonly labeled "Foundations of Silviculture," "Principles of Silviculture," or "Fundamentals of Silviculture."

Silvics: The "foundation of silviculture . . . which deals with the growth and development of single trees and other forest biota as well as of the whole forest ecosystems" (Smith et al. 1997, 3). Silvics was considered synonymous with forest ecology until the 1960s (Spurr 1964). European silviculture books (Mayer 1984; Burschel and Huss 1997; Röhrig et al. 2006) included sections describing the silvics of the main tree species. Silvics of North American tree species were described in early North American silviculture books (e.g., Toumey and Korstian 1947) and more recently by Burns and Honkala (1990).

Forest ecology had a specific purpose of "providing the foundation for silviculture" and the interpretation of ecological concepts was driven by this purpose. While the 1964 forest ecology textbook broke new ground by "considering forests as complex ecosystems" (Spurr 1964), descriptions and interpretations of ecological concepts were strongly influenced by the prevailing silvicultural viewpoints, specifically stand-scale management and the agricultural approach to research and practice (chap. 2). For example, aspects such as "lesser" vegetation, animals, and the complex biota of the forest floor and soil were dealt with only briefly. More tellingly, discussions of these aspects focused on their relationships to trees (Spurr 1964).

Early writings in forest ecology reflected the origin of the discipline and accepted and utilized the idea of homogeneous units being managed as stands. In contrast to the science of ecology, forest ecology placed less emphasis on the importance of heterogeneity at different spatial and temporal scales. The view of Kimmins (2004) that forest ecology is merely the *application* of general ecology to a forest ecosystems highlights the importance of management aspects for forest ecologists. However, even with that focus, contemporary forest ecology texts and courses now rely heavily on concepts from ecology (Perry 1994; Barnes et al. 1998; Waring and Running 1998; Frelich 2002; Kimmins 2004; Montagnini and Jordan 2005). In addition, several books have attempted to directly cover the linkage of ecological concepts to forest management (e.g., Kohm and Franklin 1997; Hunter 1999; Lindenmayer and Franklin 2002, 2003). These books promote new approaches to forestry and silviculture that have a stronger basis in concepts developed in ecology. Such developments are likely to influence the development of forest ecology as a scientific discipline per se. For example, Kimmins (2004) acknowledges that forest ecology has traditionally stressed community (or stand) levels of organization. He suggests that its emphasis should instead be expanded to all levels of biological organization within forest ecosystems.

Despite these advances, the disciplines of silviculture and ecology still view forested ecosystems in fundamentally different ways; the schism is hindering interactions between the two disciplines. To understand the reasons for this hindrance, it is useful to contrast how silviculturists and ecologists view a forest, how their professional research organizations function and interact, and how their educational materials vary. Research approaches influence the short- and long-term views of a discipline and are discussed in the final section of this chapter. Specifically, we examine why silviculture has moved to the use of large-scale experiments in response to changing societal values and new ecological knowledge. Lastly, we explore ways to increase the effectiveness of experiments that aim to address the broader objectives of managing heterogeneity of structure and ecological complexity in forests.

What Do Silviculturists and Ecologists See When They Walk into a Forest?

A walk through a forest can highlight that—admittedly, this is a caricature of real life—silviculturists and ecologists view the world through very different eyes. First, we contrast their views by "walking" through an old natural forest that has not been disturbed by human activity or large-scale natural disturbance for a long period of time.

The old natural forest may have aesthetic appeal to silviculturists, but they do not consider it desirable or productive. It is viewed as an underachiever. Silviculturists can easily imagine a "better" forest that could replace this messy and unproductive natural forest once it is under management. Silviculturists typically focus on commercial tree species and whether they are growing up to site potential. They use log and timber grading criteria as a basis to categorize individual trees as being "good" or "bad," depending on species, canopy position, tree size and shape, and general vigor. Silviculturists view disturbance agents such as windthrows, insects, and diseases as undesirable and something that can be avoided or minimized with proper management. Silviculturists believe active management will produce multiple rotations of predictable and sustainable timber.

For an ecologist, the same forest is the culmination of hundreds, thousands, or even millions (in the tropics) of years of evolution, adaptation, competition, selection, disturbance, and change. Ecologists marvel at the structural, compositional, and dynamic variability of the forest. They see a purposeful complexity in the natural forest. For example, the soil is viewed as the product of close interactions among vegetation, climate, microorganisms, and the underlying geology. Each element of heterogeneity adds to ecosystem complexity, resilience, and function. The inherent heterogeneity of the natural forest creates a variety of niches that promote or maintain biodiversity. To the ecologist, the role or function of all ecological elements and species is important. Trees receive special attention only because they are a defining feature of forests and one of the larger components of biomass, thus allowing myriad other organisms to survive and evolve. Disturbance agents are viewed by the ecologist as intrinsic components of ecosystem dynamics. In fact, some ecologists view human interventions that limit natural disturbances as the real disturbance agents (Peter Attiwill, pers. communication). Individual species are viewed in the context of adaptation to these disturbance agents and many may even require disturbance for continued survival. Change is perceived as inevitable by the ecologist. The old forest will continue to develop a variety of structures that maintain ever-changing processes and functions. The forest is viewed by the ecologist as more resilient and better able to adapt to new environmental conditions than simplified managed stands.

Table 4.1. Impact of different "lenses" for viewing forests

A temperate deciduous forest undisturbed by fire or humans	*The same forest managed by the group selection system for the past 100 years*
View of traditional silviculturists	
Uneven-aged	Uneven-aged following an inverse J curve
q-factor of 1.1	q-factor of 1.7
Overmature and decadent forest	Productive and regular forest
Basal area of 45 m^2	Basal area of 32 m^2
Lots of dead and diseased trees	Straight and healthy trees
Productivity of 1 m^3/ha/year	Productivity of 3.5 m^3/ha/year
Mixedwood cover type	Mixedwood cover type
Composed of 8 tree species	Composed of 3 tree species
An overmature, unproductive forest	A productive forest
A messy forest with gaps, crop and non-crop species, dense understory, diseases	A uniform and healthy forest
A forest that is part of the tolerant hardwood productivity group	A forest that is part of the tolerant hardwood productivity group
A forest that needs to be managed to be productive	A productive forest that plays to its full potential
View of ecologists	
An old undisturbed forest	A younger managed forest
Lots of diversity of structures and living creatures	A simplified forest that has lost lots of its structure and diversity
Beautifully large live and dead trees	Lack of large and diseased trees
Very productive forest in term of species and energy flow	Less productive forest
Composed of 13 trees, 5 shrubs, 45 herbs, 12 mosses, 65 known fungi, 4 rare-bird nests	Composed of 6 trees, 4 shrubs, 39 herbs, 11 mosses, 55 known fungi, no rare-bird nests
A nice gradient of vertical and horizontal heterogeneity	A lack of vertical and horizontal heterogeneity
One of the normal conditions for a forest of this region	A totally abnormal condition, rarely found in the region
A forest that is part of the temperate deciduous biome	A forest that is part of the temperate deciduous biome
A forest that needs to be preserved; management can only degrade the forest	A forest to restore by stopping the regular cutting or by modifying the cutting to increase the structural and functional diversity

Further detailed aspects of the impact of the different "lenses" that silviculturists and ecologists use to view forests are provided in table 4.1. In addition to the old natural forest discussed above, the table also shows that the different viewpoints are just as evident when silviculturists and ecologists walk into a managed forest. The table provides an example of how both disciplines perceive a forest that has been managed intensively by the group selection system (see Matthews 1989; chap. 1) for the last 100 years.

Who Do Silviculturists and Ecologists Talk To?

The inherent differences between the disciplines of silviculture and ecology also display themselves in the two principal organizations that oversee research in their respective disciplines. It is fairly evident that the organizational structures of the two organizations don't facilitate communication or crossdisciplinary cooperation. The International Union of Forest Research Organizations (IUFRO) is the largest forestry research organization (www.iufro.org). It divides research into eight major divisions to support researchers in collaborative work and provide an organizational link among research groups (silviculture; physiology and genetics; forest operations; forest assessment, modeling, and management; forest products; social, economic, information, and policy sciences; forest health; and forest environment). Forest ecosystems and biodiversity are mentioned under forest environment, not under silviculture, suggesting that these issues are being viewed as external to the discipline of silviculture. Within the silviculture division researchers align themselves by management objectives, practices, or geographic regions. Examples include short-rotation silviculture, forest vegetation management, even-aged silviculture, uneven-aged silviculture, and tropical silviculture. Further divisions are often based on tree species.

In contrast, the Ecological Society of America (ESA) divides research into twenty large and broad sections of interest to facilitate communication between ecologists with similar disciplinary interests (www.esa.org). Broad sections covering many terrestrial ecosystems include applied ecology, biogeoscience, long-term studies, paleoecology, physiological ecology, plant population ecology, and vegetation. Other sections are organized by study object (e.g., agroecology, aquatic, rangeland, soil, statistics, traditional ecological knowledge, and urban ecosystems) or geographic region. Forests are one of the most important terrestrial ecosystems being studied by ecologists, but no direct mention is made of forests or trees in the research structure of the ESA.

Only a small number of ecologists attend meetings sponsored by IUFRO, and very few silviculturists regularly attend ecology meetings sponsored by the ESA or INTECOL (International Association for Ecology). The timing of their annual meetings in 2005 is reflective of this pattern in attendance. The ESA/INTECOL meeting in Montreal, Canada, and the IUFRO Congress in Brisbane, Australia, were held at exactly the same time. The lack of coordination in scheduling meetings, in conjunction with different organizational structures and often different physical and administrative locations in universities, does not encourage exchanges between silviculturists and ecologists.

The two disciplines are slowly beginning to overcome this separation. More and more professors in forestry schools have a biology or ecology background, but the reverse is rare. This pattern can be explained by differences in research approaches typically found in forestry schools and biology and ecology departments. Doctoral students in forestry schools are more likely to study management-oriented issues and their theses often do not address basic or theoretical questions. In contrast, in biology and ecology departments, it is common for PhD theses to cover basic and theoretical topics. Accordingly, forestry students more often publish in forestry journals, which generally have lower impact factors (number of times a paper is cited per two years) than major ecological journals. Since the amount and quality of publications are foremost criteria for many hires, applicants with a PhD from a forestry program have a difficult time obtaining work in biology and ecology departments. Forestry schools are not immune to pressures to publish in prestigious journals and use this in their hiring procedures as well. Consequently, students from ecology and biology departments with publications in journals with high impact factors are of high interest for positions in forestry schools. However, they often struggle with the reputation of being too theoretical to contribute to management-oriented forestry programs.

What Do Silviculturists and Ecologists Read?

Educational materials, especially textbooks, are a defining feature for any discipline. A closer look at commonly used textbooks in silviculture and ecology provides further insight into how and why the two disciplines view forests so differently. In North America, silviculturists attain knowledge about forest ecosystems from silviculture and forest ecology courses, which commonly use textbooks such as Smith et al. (1997) or Nyland (2002) for silviculture and Kimmins (2004) or Barnes et al. (1998) for forest ecology. In contrast, ecologists learn about forest ecosystems from general population, community, and ecosystem ecology textbooks, such as Begon et al. (2006). A comparison of headings and subheadings in these three groupings of textbooks provides further appreciation of the different views of silviculturists and ecologists (tab. 4.2).

Table 4.2. Headings and subheadings used in silviculture, forest ecology, and general ecology textbooks

Silviculture	*Forest Ecology*	*Ecology*
Smith et al. 1997	Kimmins 2003; Barnes et al. 1998	Begon et al. 2006
Focus on forests, trees, and stand averages	Focus on forests and trees	Focus on vegetation community
	Topics of divisions and/or chapters	
Stand dynamics	Ecosystem concepts	Organisms and environment
Various types of cutting	Production ecology	Environmental conditions
Ecology of regeneration	Biogeochemistry (soil)	Resources
Site preparation	Ecosystem classification	Demography
Planting	Physical environment	Migration and dispersal
Site classification	Population and community	Interactions
Stand development	Genetic and evolution	Behavior
Even and uneven stands	Temporal diversity	Communities
Mixed stands	Spatial diversity	Flux of energy
Damaging agents	Environmental issues	Community structure
Wildlife habitats	Models	Disturbance
	Sustainability	Island biogeography
		Complexity, stability, and structure
		Pattern of species richness

Concepts and Theories Can Provide a Linkage between Silviculture and Ecology

The previous sections may have given the impression that silviculture and ecology are isolated, but the two disciplines do not exist in a vacuum. Ideally, they should complement each other. Linkage between the two disciplines is very important, as they have a lot to learn from each other. For example, the emerging concepts around ecosystem resilience and function are based on information provided by general ecology, yet are critical to the silvicultural management of forests. Alternatively, forests have been studied, managed, and monitored by silviculturists for a long time, which provides unique opportunities to evaluate and test basic ecological theories.

In chapter 3, we listed selected ecology concepts and their counterparts in silviculture (tab. 3.1). Despite an apparent linkage of associated concepts, the two disciplines still often view individual concepts in the context of their respective disciplinary boundaries. The interpretation and use of the niche theory, especially how the two disciplines relate the niche theory to variability, provide an example of the impact of disciplinary boundaries (fig. 4.1). Silviculturists use the niche concept as a tool to aid decisions about which tree species to regenerate in managed forests. For reasons of efficiency and predictability, silviculturists often establish a desired tree species over as wide a range of sites as possible; that is, across most of the fundamental niche of the species. In contrast, ecologists view the niche concept as a tool to understand how nature works. Ecologists may look at the same graph, but are more likely to focus on the multitude of species that appear to overlap in terms of their fundamental niches (fig. 4.1).

The Sub-Boreal Spruce forests (SBS, Meidinger and Pojar 1991) that dominate the landscape of the central interior of British Columbia, Canada, provide an illustration of the different interpretations of the niche theory. The SBS forests cover approximately 103,000 square kilometers, equivalent in area to Iceland or nearly the combined size of Belgium, Netherlands, and Switzerland. The forests established mostly after natural fires, and forest management is just beginning to have an impact in the region. Natural forests in this region are typically composed of up

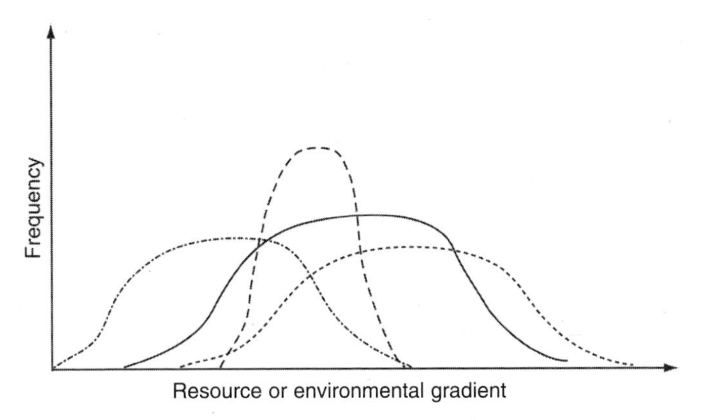

Figure 4.1. Conceptual niche distribution of multiple species along a resource or environmental gradient.

to eight tree species. Mature forests are frequently dominated by lodgepole pine (*Pinus contorta*), interior spruce (*Picea glauca x engelmannii*), subalpine fir (*Abies lasiocarpa*), or Douglas fir (*Pseudotsuga menziesii*). Trembling aspen (*Populus tremuloides*), paper birch (*Betula papyrifera*), black cottonwood (*Populus trichocarpa*), and black spruce (*Picea mariana*) are also present across significant portions of the landscape. Current silvicultural practices after harvesting, as described in forest stewardship plans, call for the planting of two preferred species, interior spruce or lodgepole pine, across virtually all site types in the SBS forests. Exceptions include the occasional plantings of Douglas fir on specific site types. Thus, the majority of the naturally occurring tree species are not favored for planting. Furthermore, silvicultural activities are actively discouraging them, if their natural regeneration makes up more than 20 percent of the stand. The planting of lodgepole pine and interior spruce on virtually all site types reflects a silvicultural focus on the wide fundamental niche of these species. As a result, silvicultural practices reduce the variability within the large area of SBS forests by promoting only two tree species. A contrasting ecological interpretation of the niches in these forests would focus on variability in species patterns in the landscape and conclude that the SBS forests can easily support multiple tree species in a diverse array of mixtures.

Interpretations of issues such as the pros and cons of single- versus mixed-species forests provide another example of the different disciplinary viewpoints. Silviculturists and ecologists have been interested for years in ecosystem productivity. Numerous studies and publications have dealt with the question of whether mixed-species stands (silviculturists) or biodiversity (ecologists) optimizes yield (silviculturists) or ecosystem functioning (ecologists). As described by Pretzsch (2005), silviculturists have debated the virtues of single-species and multispecies stands at least since Hartig produced his classic forestry science text (Hartig 1791). For silviculturists, the external factors described in chapter 1 are clearly influencing their view. For example, stands with single or few tree species are often preferred because of higher management efficiency. Other aspects, such as concerns about biodiversity, don't raise sufficient concerns by landowners and silviculturists to override the efficiency argument. Furthermore, predictive tools (yield tables, models) were, until recently, mostly developed for single-tree species stands (chap. 2). Thus, single-species stands were also preferred by silviculturists because of their better predictability.

In contemporary ecology, the relationship between biodiversity and ecosystem functioning is undergoing a major research thrust (see Loreau et al. 2002; Hooper et al. 2005). Parallel to silviculturists, ecologists are studying whether the biota (e.g., species composition) affects ecosystem function, measured for example as biomass production (Tilman et al. 2002a). However, ecologists discuss these issues in terms of unifying principles and theories, such as "niche complementarity" or the "insurance hypothesis" (Loreau et al. 2002). For reasons of research efficiency, their experimental studies to test these theories are typically not utilizing trees, but are often implemented over very short time intervals or with short-lived species. Consequently, their short-term experiments provide few direct insights into the practical aspects of the management of single-species versus multispecies forests.

Silvicultural studies are rarely based on theories, nor do they necessarily seek to determine the mechanisms behind single- versus multi-species productivity. Silvicultural investigations of productivity and species mixture are strongly influenced by the availability of local long-term datasets from permanent sample plots (e.g., Pretzsch 2005). Although there are many empirical datasets available, the unfortunate result of relying mostly on empirical analysis without a strong theoretical basis is often a lack of generalization. The unique characteristics of each sampled forest will impose significant constraints on generality. This may explain why silviculturists continue to debate the yield implications of mixed-species and monocultures with seemingly no resolution. The analysis of long-term datasets by Pretzsch (2005) is good example of an attempt to resolve the conflict between unifying principles and site-specific management based on the individual strengths of silviculture and ecology.

Obviously, both disciplines would benefit greatly by closer collaboration around many important issues. While it is generally accepted that ecological principles can and should serve as the primary basis for management of natural ecosystems (McPherson and DeStefano 2003), conceptual linkages between the two disciplines are often still lacking.

The Evolution of Contemporary Large-Scale Silvicultural Experiments

During the 1980s and 1990s public perceptions about forests rapidly evolved, bringing pressure on silviculturists to manage forests for a variety of ecological, social, and economic goals. In response, researchers from many disciplines paid increasing attention to the role of structure and disturbance in maintaining biodiversity and resilience in forests. Research into the nature and role of old-growth characteristics on ecosystem processes and population or community dynamics in forests also became a major topic of interest, especially in regions with substantial remaining natural forests.

Wildlife, Forests, and Forestry by Hunter (1990) provides an excellent introduction to the importance of structure in forested ecosystems in the context of wildlife habitat. Other aspects were covered subsequently in other papers and books. For example, Bunnell et al. (1999) reviewed structural management tactics to maintain vertebrate richness in managed stands and highlighted the importance of different tree species, trees of varying size, dead and dying trees, downed wood, shrubs, and riparian areas. The role of old-growth forests and the implications of varying frequency, intensity, and pattern of disturbance on innumerable aspects of forested ecosystems received extensive study worldwide. Of particular interest to silviculturists are books and reviews by Platt and Strong 1989; Attiwill 1994; Denslow and Hartshorn 1994; Fries et al. 1997; Angelstam 1998; Hunter 1999; Franklin et al. 2000; Lindenmayer and Franklin 2002; Bergeron et al. 2002; Burton et al. 2003; and Perera 2004.

Emerging from this collective body of research has been a clearer understanding of which structural components are vital for specific habitats, how forested ecosystems recover from disturbance, the role of legacies (cf., Franklin et al. 2000) in ecosystem recovery, and more generally, how forested ecosystems maintain resilience (e.g., Folke et al. 2004; Drever et al. 2006). These developments are of considerable importance to the practice of silviculture. Accordingly, new silvicultural practices, mostly involving structural retention by leaving various amounts and patterns of live and dead trees at the time of harvest, have been proposed and implemented by silviculturists and forest ecologists (Seymour and Hunter 1992, 1999; Kuuluvainen 1994, 2002; Coates and Steventon 1995; Bergeron and Harvey 1997; Coates and Burton 1997; Vanha-Majamaa and Jalonen 2001; Franklin et al. 2002; Harvey et al. 2002; Palik et al. 2002; Seymour et al. 2002; Lieffers et al. 2003; Kangur 2004; Seymour 2005).

Silviculturists, especially those working on public lands, were challenged to reevaluate or defend their traditional practices (discussed in chap. 2) or develop new practices in response to these pressures. As they tackled this challenge it quickly became obvious that traditional research methods could not address the variety of questions being asked. Responding to these challenges required a different approach to research. In response, silviculturists installed new collaborative large-scale management experiments throughout the 1990s and into the early 2000s. Because many of the questions were centered on ecological responses of interest, experiments with larger spatial and temporal scales than typically applied in "small-plot" agricultural-style silvicultural experiments were required (Ganio and Puettmann, 2008). These experiments were established at operational scales that minimized the need to scale up from small research plots to operational stands. It also allowed measurement of a broad range of response variables to characterize the response of numerous aspects of ecosystem development to the experimental treatments, most of which focused on amount and spatial distribution of retention trees.

These studies were new, and provided the exciting possibility of quantitatively linking ecological theory to silvicultural practices. They were labeled Large-Scale Management or Silvicultural Experiments, Alternate Silvicultural Systems Experiments, or Emulating Natural Disturbance Experiments. Major examples in North America are MASS (Arnott and Beese 1997), Date Creek (Coates et al. 1997), DEMO (Aubry et al. 1999), EMEND (Spence et al. 1999), Sicamous Creek (Vyse 1999), SAFE (Brais et al. 2004), OMEM (Guldin 2004), SOYDF and STEMS (Curtis et al. 2004; de Montigny 2004), RSCP (Palik et al. 2005) and AFERP (Seymour 2005). Several of these and other multidisciplinary experiments are reviewed in Monserud (2002), Peterson and Maguire (2005), Seymour et al. (2006), Kuehne and Puettmann (2006), and Poage and Anderson (2007). These new experiments were a direct response to changing social and ecological views of forests and their management. Of course, several older silvicultural experiments also can provide insight into some of the pressing questions of today. Seymour et al. (2006) provide an excellent review of this topic for U.S. forests.

The overall objective of the contemporary large-scale silvicultural experiments was to investigate new options for incorporation of greater structural and ecological heterogeneity into current silvicultural practices. These experiments therefore provided opportunities to investigate forest ecosystem responses at a variety of spatial and temporal scales. The large size of treatment areas, often 5 to 30 hectares in size, also allows for assessment of treatments designed to create small-scale variation and within-stand diversity. Treatments that represent different levels or patterns of structural retention in conjunction with different sizes and shapes of openings for regeneration of new trees were of special interest in many of these studies (e.g., Fahey and Puettmann 2007). Despite these innovations, the experiments still show their roots in the agricultural research model and associated statistical procedures that seek and value uniformity and stand-scale application (see chap. 2).

Seymour et al. (2006, 106) write succinctly in their review of four large-scale experiments in the United States that "all studies use the time-tested randomized complete block design with all treatments represented in a single location." Virtually all contemporary large-scale silvicultural experiments mentioned previously rely on agricultural experimental designs and standard parametric statistical tests. While providing statistical strength for addressing some of the questions being asked, the continued reliance of silviculture on these statistical approaches results in unique challenges when trying to understand a variety of ecosystem responses at various spatial and temporal scales. Especially challenging were investigations of the impacts of different patterns of structural retention and removal that occur at smaller scales than the treatment-unit scale. The silvicultural emphasis on treatment-unit scale analysis of mean responses, even when the rationale for the treatment was to increase within treatment structural variability (Monserud 2002; Kuehne and Puettmann 2006), can pose considerable difficulties.

Many individual treatments are a composite of smaller-scale manipulations, such as a thinned forest matrix, cut gaps, and leave tree islands (e.g., Cissel et al. 2006). An understanding of such small-scale variability cannot be achieved efficiently in agricultural experimental designs (chap. 2; Ganio and Puettmann 2008). Experiments following the agricultural model work best when variability is tightly controlled. The discrepancy between experimental design and treatment rationale can lead to concerns that any results from structurally variable treatments in these experiments are problematic because of the "low precision owing to the coarse scale." This quote, from an anonymous associate editor, highlights the inherent conflict between the desire of researchers to investigate within-stand variability and problems with high variability when analyzing data from studies that are based on an agricultural research model. In addition, the value of representing the overall treatment condition with a single value, usually the mean, becomes questionable. As discussed in chapter 2, the average is less informative when it is representing areas that were purposely treated to be highly variable, such as applied in many of the large-scale silvicultural experiments.

One of the greatest challenges regarding large-scale silvicultural experiments is the decision about where to establish sample plots that represent the experimental treatment, especially in large structurally variable treatment units (Ganio and Puettmann 2008). Furthermore, from an experimental perspective, it is not always obvious what exactly the treatment is in contemporary large-scale silvicultural experiments. We illustrate these points using the Date Creek Silvicultural Systems Experiment as an example.

The Date Creek Silvicultural Systems Experiment and Alternate Study Methods

The Date Creek Silvicultural Systems Experiment is a multidisciplinary set of individual studies examining tree growth and ecosystem responses in the transitional coastal-interior forests of northwestern British Columbia, Canada (Coates et al. 1997). The overall experimental design is thoroughly described in several individual studies (Coates 1997, 2000, 2002; Steventon et al. 1998). Like other large-scale silvicultural experiments, the Date Creek experiment employed a randomized block design to organize four replicates of four different structural retention treatments, for a total of sixteen individual treatment units. Each treatment unit was about 20 hectares in size. The four treatments were no removal (the undisturbed forest), light and heavy partial cutting, and clearcutting. In the light partial cutting, about 30 percent of the stand volume was removed by cutting either single stems or small gaps (3 to 10 trees). In the heavy partial cutting, about 60 percent of stand volume was removed. Here, the cutting pattern used both large gaps (500 to 5,000 square meters in size), evenly distributed across the treatment units, and either single-tree or small gaps (less than 300 square meters) in the forest matrix between the larger gaps.

The first challenge was to decide where to place sampling plots in each treatment unit to properly represent the conditions found in that treatment. Recall from chapter 2 that designed experiments use plot designs that average across variation. Because all treatments, except the clearcut, contained variable-sized patches of retained trees and gaps (cut or natural) in different spatial patterns, the within-treatment-unit variability was often as great as or greater than the variability among treatment units. Standard sampling procedures include randomly or systematically placing equal numbers of sample plots in each treatment unit. Unless an unreasonably large number of plots can be established, sample plots will very likely not be representative of the suite of conditions found within the treatments. Many plots would probably fall in similar conditions within and among treatment units. For example, extended areas within the 30 percent removal treatment, the matrix of the 60 percent removal treatment, and the undisturbed forest treatment all have similar tree densities. There is also overlap in gap opening size among these three treatment units. The gaps make up a relatively small portion of each treatment unit, but with very different conditions from the remainder of the treatment unit. With a random or systematic sampling scheme, these areas are likely sampled insufficiently to allow a solid description of the conditions. Also, some research objectives are interested in smaller-scale processes, for which it may be important to test whether sections of large gaps in the heavy removal treatment are similar in condition to the clearcut treatment unit. This and similar questions highlight that a "simple" decision as to where to place sampling plots is only one of the many inherent challenges faced when investigating aspects of structural and process variability within experiments that utilize experimental designs based on the agricultural research model (Ganio and Puettmann 2008).

The Date Creek experiment also demonstrates that researchers can gain a lot of insight when thinking outside the box of agricultural experimental designs. At Date Creek, individual studies were designed to operate at one of three scales: the microsite, the gap, and the treatment unit scale (Coates et al. 1997). As it turned out, very few questions actually were appropriately studied at the treatment unit scale; that is, directly utilizing the overall agricultural experimental design of the experiment. Most questions that were of interest for developing new and innovative silvicultural treatments were appropriately addressed using the variability created within and among the treatments to study response variables (e.g., tree growth) across gradients of conditions (e.g., light levels, Coates and Burton 1999; gap size, Coates 2000; or as a function of the composition and abundance of the local tree neighborhood, Canham et al. 2004), or under particular conditions (e.g., ectomycorrhizal mushroom response, Kranabetter and Kroeger 2001).

Rather than being bound to comparing a limited set of treatments in search of a best treatment, viewing treatments (or better yet, establishing treatments) as a means to provide a gradient of contrasting conditions for the study of specific silvicultural questions is a far more productive approach in such settings. This approach allows a better understanding of ecosystem responses to various patterns of structural retention and removal typically applied in large-scale silvicultural experiments. First, plants and other ecosystem components actually respond to the conditions created by a treatment, and not the treatment itself. Second, adapting the spatial scale used to address particular questions to the spatial scale of the process of interest will greatly improve the results coming from large-scale silvicultural experiments (e.g., LePage et al. 2000).

New Analytical Tools Can Help

The debate on how forests should be managed and how researchers can best help guide this debate will continue. Studies that investigate aspects of scales and scaling will be of special importance in this context. For example, it is now becoming well understood that interactions among individual trees and their spatially heterogeneous environment are inherently local in nature, acting at a neighborhood scale over restricted distances (Stoll and Weiner 2000; D'Amato and Puettmann 2004; Gratzer et al. 2004; Canham and Uriarte 2006). This concept is very useful when trying to understand and/or manage small-scale variability in forest structures and processes. In forests, the spatial distribution of canopy tree species can exert a strong control over the interactions of other organisms and ecological processes, all with possible feedbacks that in turn can influence canopy tree dynamics (Canham and Uriarte 2006 and references therein). Remember, silviculturists manage the establishment, survival, and growth of trees and all these demographic processes unfold at local neighborhood scales.

The study of forest dynamics and, more specifically, the study of individual tree neighborhood dynamics is particularly well suited to the use of likelihood methods and model selection techniques (Hilborn and Mangel 1997; Burnham and Anderson 2002; Johnson and Omland 2004; Canham and Uriarte 2006; Hobbs and Hilborn 2006). At the heart of the methods is the explicit interplay between data and models, with "model" used in the sense of a mathematical statement of the quantitative relationships that are assumed to have generated the observed data

(Canham and Uriarte 2006). Classical hypothesis testing (see chap. 2) is replaced by the more general process of model selection and comparison, using likelihood and parsimony to compare the strength of evidence for competing hypotheses, represented by the different possible mathematical models (Johnson and Omland 2004; Canham and Uriarte 2006).

Model Selection: Technique that emphasizes evaluation of the weight of evidence for multiple hypotheses by seeking accurate and precise estimates of parameters of interest, for example, factors affecting understory tree growth. Model selection evaluates competing hypotheses against observed data and aids identification of the mechanisms most likely to explain tree growth as a function of local neighborhood conditions. Traditionally, models used by silvicultural researchers were limited to a relatively small set of linear forms that did not explicitly represent biological states and processes (Hobbs and Hilborn 2006). Model selection has three primary advantages over null hypothesis testing (Johnson and Omland 2004): (1) it is not restricted to a single model, measured against some arbitrary probability threshold; rather, multiple models are assessed by comparing relative support in the observed data; (2) models can be ranked, thus providing a measure of support for each hypothesis; and (3) if competing hypotheses have similar levels of support, model averaging can be used to make robust parameter estimates and predictions. Dramatic increases in computer power have made it far easier to use these techniques than it was in the past.

Likelihood methods have several advantages. They provide analogues for many traditional parametric statistical tests, but often without many of the restrictive assumptions required for parametric statistics (Canham and Uriarte 2006) that can cause such problems in the analysis of the traditional experimental designs used in the large-scale silvicultural experiments. Another strong advantage of likelihood methods and model selection is the ability to easily link these methods to the development of spatially explicit, individual-based models that can capture interactions among individual organisms, thus encapsulating the theory of neighborhood dynamics (Gratzer et al. 2004). Results from a typical analysis of variance with a fixed small number of different treatments are generally difficult to incorporate into dynamic models.

Likelihood methods can provide powerful tools for hypothesis testing via model selection. The effectiveness of the approach, though, ultimately lies in the insight of the investigators in choosing appropriate and interesting scientific models and their skills in collecting appropriate data (Canham and Uriarte 2006).

Likelihood methods and model selection techniques are still seldom used in silviculture (but see Kobe and Coates 1997; LePage et al. 2000; Canham et al. 2004; D'Amato and Puettmann 2004) due to the strong dominance of the null hypothesis testing and the search for a best treatment that still dominates the discipline. Ecological researchers are grappling with similar issues of adapting their research and analytical approaches to changing research questions. They have also relied on agricultural experimental designs and are finding these methods inadequate for addressing many contemporary ecological questions (Hobbs et al. 2006). However, ecologists have had a longer interest in understanding heterogeneity in nature (see chap. 3) that has forced them to be more open to new statistical approaches that can better address aspects of scale and variability. Ecologists may be more advanced than silviculturists in the use of innovative statistical techniques, and their experiences can be of great benefit to silviculturists.

Conclusion

Ecologists and silviculturists have different niches in the management of natural resources and consequently have developed different views of forested ecosystems. Disciplinary differences exhibit themselves in a variety of settings, such as the structure of research organizations and different interpretation of concepts. The historic disparity in goals between the two disciplines has inhibited communication and coordination and impaired collaboration, yet both are deeply concerned about the viability of forested ecosystems.

The two disciplines now share a similar desire to understand the important processes driving productivity and resilience in forests. The recent establishment of a new series of large-scale silvicultural experiments to address silvicultural and ecological questions of management interest is a promising development and should help close the historical gap between the disciplines. Most importantly, these experiments show the benefits that can be gained when researchers work together in studying heterogeneity in forests. Large-scale silvicultural experiments aimed at developing new, innovative management practices are still dominated by the agricultural research model. This imposes considerable constraints on studying heterogeneity and complexity in forests and on identifying the important mechanisms controlling productivity and resilience. Silviculturists need to adopt techniques and consider new conceptual frameworks that can better address the challenges of the twenty-first century.

5

Managing Forests as Complex Adaptive Systems

The societal view of the role and importance of forests and the methods used to manage forests has undergone recent changes in many regions of the world. These developments are especially prevalent on public forest lands. Increased public and professional concerns about the maintenance of biodiversity and the disappearance of primary forest all over the world are forcing silviculturists to acknowledge and accommodate a wider range of ecological and social issues than ever before. Ecological research over the last few decades has also increased our understanding of ecosystem functions and processes and how they are affected by natural and managed disturbances (see chaps. 3 and 4). Foresters in general, and silviculturists in particular, are under pressure to respond to this paradigm shift (see chaps. 1 and 2).

The well-established and long-held traditions of silviculture are generally viewed as the strength of the discipline, but they are also proving to be obstacles as silviculturists are faced with wider varieties of management objectives and constraints. We suggest that the discipline of silviculture will benefit from a new conceptual framework that will aid silviculturists in addressing present-day forest management issues. Silviculture is at risk of becoming marginalized in broader forest management decisions if it does not respond to societal expectations and new ecological information. It is often perceived narrowly as being capable only of growing trees for timber production. The historical success of silviculture in improving timber yields has become a liability as the focus of management has shifted to broader issues such as sustaining the full function and dynamics of forested ecosystems, maintaining biodiversity and ecological resilience, and providing for a variety of ecosystem services of value to humanity.

We suggest that these challenges can be addressed by setting a new path for silviculture, which we label "managing forest as complex adaptive systems," that will benefit from embracing ecological viewpoints and approaches better suited to deal with ecosystem complexity, variability, unpredictability, and adaptability. The earlier chapters of this book argue that this path requires much more than selecting different practices or silvicultural systems from an established and evolving pool of practices. While newly evolved practices may be a step forward in meeting current and future demands in forestry, managing forests as complex adaptive systems requires a major shift in philosophical and research approaches, new management tools, and a new conceptual framework to organize thinking within the discipline. We believe silviculture will remain an influential discipline that will guide future management of forest ecosystems if this challenge is tackled with a critical and open mind.

Arguing that silviculturists should manage forests as complex adaptive systems is the *raison d'être* of this book. In this final chapter we encourage silviculturists to consider ideas about complex systems in their management efforts by (1) explaining the science of complexity and its uses in many fields, (2) showing why forest ecosystems should be considered and managed as complex systems, (3) comparing the impacts of traditional silviculture on the stand structure and adaptive potential of three contrasting forest ecosystems, (4) providing a short discussion on how and which characteristics of complex systems can be incorporated into silvicultural management and research, and (5) highlighting selected management practices to move toward a silviculture aimed at managing forests as complex adaptive systems.

The Science of Complexity

The science of complexity has a varied history in multiple fields (for an overview see Waldrop 1992). One of the first scientists to come face to face with complexity was the French mathematician Henri Poincaré (1854–1912). His attempts to find a solution to such a "simple" problem as predicting the orbits of three planets that interact in a nonlinear fashion provided important concepts that influenced the chaos theory nearly a century later. Starting in the 1940s with fairly simple investigations of complex interactions among mechanical parts, the science of complexity expanded toward investigating both living and man-made systems (Weaver 1948). Early efforts focused around aspects of control, communication, adaptation, and hierarchy (Delic and Dum 2006). The disciplines of physics and economics were the most active in developing the science of complexity.

A seminal paper by Anderson (1972) introduced the idea of emergent properties, suggesting complexity science as an alternative to reductionist science. Other milestones included investigations into randomness and scales that, among others, led to the development of the chaos theory (May 1974; see also chap. 3) and the concept of fractal geometry (Mandelbrot 1977). This progress was made possible because of the advent of computers and their subsequent rapid increase in computing power (Emmeche 1997). A major development in the science of complexity was the founding of the Santa Fe Institute in 1984 as an independent research and education center whose main goal is furthering the

---

Reductionist Science: Studies objects by breaking them down into their individual parts. It relies on the assumption that the functioning of the whole system is equal to the sum of its parts. In contrast, complexity science suggests that many systems cannot be understood by looking only at the individual parts. Interactions among the system's component parts give rise to emergent properties that are much more than the sum of their parts. Intelligence is one such emergent property that cannot be explained by looking solely at the individual neurons. Ecological resilience is an example of an emergent property of ecosystems that cannot be predicted by studying the individual parts of the system.

---

understanding of complex systems through interdisciplinary collaborations in the physical, biological, computational, and social sciences (Waldrop 1992). The science of complexity is not a discipline per se, but a set of theoretical frameworks that apply to systems in a wide variety of fields including environmental, technological, biological, economic, and political problems and challenges.

"Complex Adaptive Systems" are defined as complex systems in which the individual components are constantly reacting to one another, thus continually modifying the system and allowing it to adapt to altered conditions (Levin 1998). While somewhat related to Darwinism, adaptation in ecosystems differs from evolution in species in that adaptation is acting at the level of individual components, and not the level of the system itself (Levin 2005). Arthur's (1999) theory of innovations is a prime example of the power of the complexity approach and one of the first examples of its potential usefulness to business and economics. The theory utilized feedback loops and nonlinear dynamics to better predict innovations as they relate to business cycles and provided a major advancement over classical (equilibrium-based) economics (Arthur 1999; Delic and Dum 2006). Since then, complexity science has invaded many fields and shown practical applications in business, economics, the social sciences, climatology, transport, and neurology (Delic and Dum 2006).

Ecology has always been viewed as the science of understanding the diversity of nature (see chap. 3). Interestingly, however, the application of complexity theory to ecology and biology (or biocomplexity) is relatively recent (Levin 1998, 2005; Naeem 2002; Folke et al. 2004; Solé and Bascompte 2006). The study of ecosystems as complex adaptive systems investigates how systems such as forests are organized, how relationships among individual parts or processes can give rise to collective behaviors that cannot be readily predicted by looking only at individual parts (i.e., emergent properties), and how the system adjusts and adapts to changing conditions.

Any biological system can be classified as complex and adaptive (fig. 5.1) if it displays the following properties: (1) it is composed of many parts (trees, insects, soil, and so on) and processes (nutrient cycling, seed dispersion, tree mortality, decay, and so on); (2) these parts and processes interact with each other and with the external environment in many

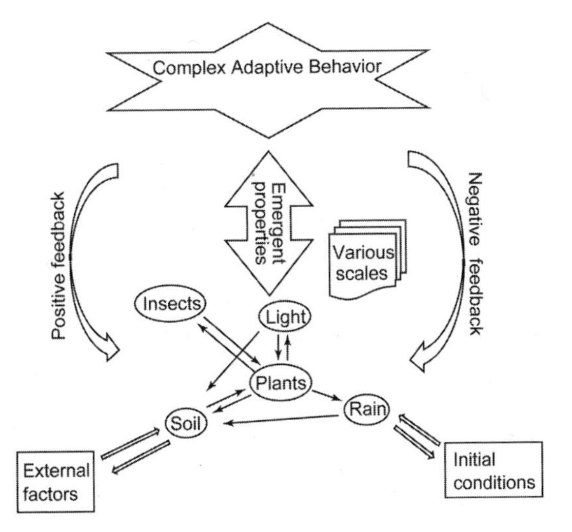

Figure 5.1. Simplified graphical representation of forest ecosystems as complex adaptive systems. Adapted from http://en.wikipedia.org/wiki/Image:Complex-adaptive-system.jpg. Accessed January 9, 2008.

different ways and over multiple spatial and temporal scales; (3) these interactions give rise to heterogeneous structures and nonlinear relationships; (4) these structures and relationships are neither completely random nor entirely deterministic, but instead represent a combination of randomness and order; (5) they contain both negative and positive feedback mechanisms, stabilizing or destabilizing the system, depending on conditions; (6) the system is open to the outside world, exchanging energy, materials, and/or information; (7) it is sensitive to the initial conditions following a major disturbance and subsequent perturbations; and (8) it contains many adaptive components and subsystems nested within each other, giving rise to emergent properties.

Forests as Complex Adaptive Systems

Among biological systems, forests could be considered the poster child of complexity and yet, the implications of this have not been directly considered by silviculturists (Folke et al. 2004). Forests contain thousands of interacting species and ecological processes, with their myriad soil organisms, herbs, lichens, mosses, insects, birds, and mammals that live and interact with each other and their outside environment across multiple spatial and temporal scales. Forested ecosystems can modify themselves (i.e., adapt) in response to their environmental and biological surroundings. Small differences in starting conditions and in nonlinear feedback loops can result in large and unexpected differences in the development of complex systems (May 1974; Solé and Bascompte 2006). Complexity science suggests that all aspects of forest ecosystems may never be highly predictable. While qualitative forecasts may be possible, the precise quantitative prediction of attributes such as total biomass, composition, or structure may pose insurmountable challenges.

It should also be evident that even the most homogeneous, intensively managed mono-specific tree plantations or intensively managed uneven-aged forests possess many attributes of a complex adaptive system. They have a natural tendency to adapt and without continued top-down management control will likely change and deviate substantially from the originally intended condition, especially after unexpected disturbance events.

To fully appreciate this new view of forests requires an understanding of how complexity in forests develops and operates. In the remainder of this section, we will further explain some key characteristics of complex adaptive systems such as nonlinear relationships, feedbacks, emergent properties, and adaptability.

Nonlinear Relationships

Nonlinear relationships occur when one variable affects another in a disproportionate way. Many such relationships exist in forest ecosystems, including Michaelis-Menten (i.e., saturation) curves for nutrient uptake, exponential or logistic population growth, and the normal, skewed, or bimodal distribution of species along environmental gradients. Many nonlinear relationships are monotonic, that is, they simply increase or decrease over the range of response of a given variable, or they may be nonmonotonic, increasing over parts of the range and decreasing over other parts (fig. 5.2).

Nonmonotonic relationships may also include threshold values where the effects of one variable or process on another can suddenly

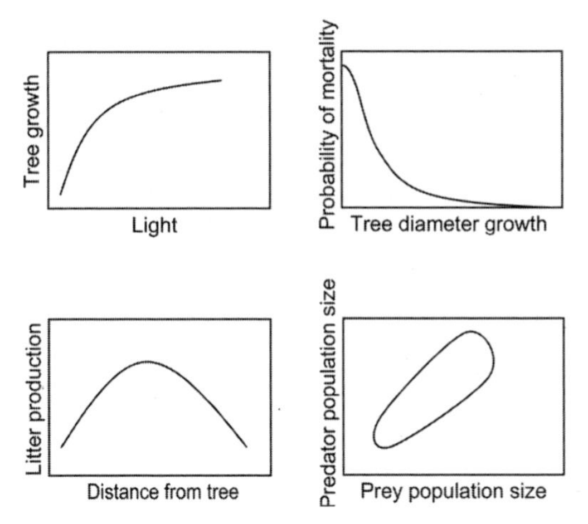

Figure 5.2. Conceptual examples highlighting nonlinear relationships between parameters as found in forest ecosystems.

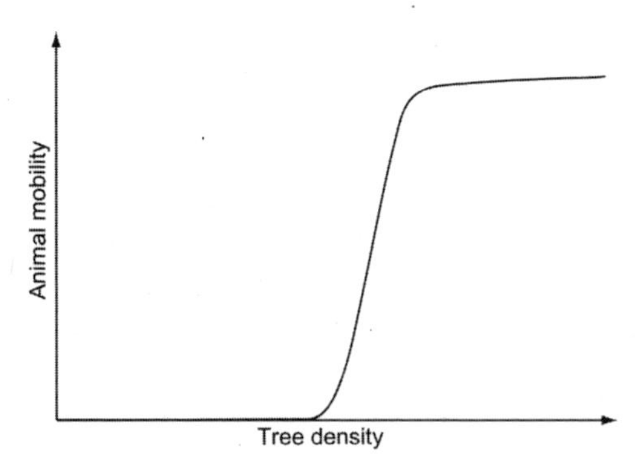

Figure 5.3. Conceptual example of a threshold relationship in forest ecosystems. Animals that need a minimum distance between tree crowns to travel through a forest have limited mobility until a critical tree density is reached. After tree density is sufficient to allow travel through tree crowns, further increases in density do not influence the animal's mobility.

start to have a much greater impact (Walker and Meyers 2004). Examples of nonlinear relationships that show threshold patterns include aspects of landscape or forest fragmentation (Green et al. 2005). Figure 5.3 shows the probability that a mammal, limited in its movement to tree crowns, is able to cross a forest. Assuming somewhat regular tree spacing, increases in tree density have no effect on the species' mobility until a minimum density level is reached. Over a narrow range of tree densities, there is a strong positive relationship between tree density and animal mobility. However, once the density is high enough to allow the animal to move across the forest, further increases in tree density have no effect.

Feedbacks

Positive or negative feedback mechanisms are also common in forests. Positive feedbacks occur when an increase in input to a system leads to an increase in output, whereas negative feedbacks occur when an increase in input leads to a decrease in output. Positive feedback loops tend to destabilize systems because they accelerate or amplify changes in system states; negative feedback loops stabilize systems because they tend to inhibit or dampen changes. Examples of positive feedback loops include the tendency of tree species that possess adaptations such as serotinous cones, which enable them to take advantage of conditions following fires, to alter their environment through fuel accumulation so as to create flammable conditions that encourage more fires (Schwilk 2003). A more dramatic example is the link between global warming and the melting of permafrost soils. As global warming increases, the melting will accelerate,

---

Lodgepole Pine: Forests in the interior northwest of North America are an example of an ecosystem that is highly homogeneous in tree species makeup and stand structure and at the same time well adapted to be resilient to recurrent wildfire. Lodgepole pine's serotinous cones allow it to establish at high densities after a fire, thus choking out other tree species that would otherwise outgrow and overtop the pines. Lodgepole pines typically die young, leaving a high fuel load that generates catastrophic fires that in turn favor lodgepole regeneration. However, when fires are suppressed the system becomes unstable and lodgepole pine may be reduced in abundance.

---

which in turn will accelerate greenhouse gas emissions from soils. A third example is the favorable understory environment that hemlock trees produce to facilitate regeneration of their own seedlings (Catovsky and Bazzaz 2000). The predator-prey relationship is a classic example of a negative feedback where the exponential growth of a prey population is inhibited by the growth response of the predator population (Roughgarden 1979). Negative feedback loops also commonly occur in forests when initially high growth rates of trees after a disturbance become progressively slower as nitrogen is immobilized in plant biomass and on the forest floor and competition for light intensifies.

Emergence

Emergent properties are system-level phenomena that cannot be easily observed or predicted by studying the individual parts of the system (Ponge 2005). Most obviously, trees themselves and tree growth can be thought of as emergent entities as their structures, functions, and processes cannot be predicted by detailed understanding of their individual cells or tissues. At a larger scale many forest insect and disease outbreaks, such as the recent unprecedented mountain pine beetle and Dothistroma needle blight epidemics in British Columbia (Carroll et al. 2004; Woods et al. 2005), are examples of emergent phenomena that result from cross-scale, nonlinear interactions among the insect or pathogen, the host tree, stand- and landscape-scale forest practices, and climate conditions. Some would say that self-organization, resilience, and adaptability of forest ecosystems are their most important emergent properties (Gunderson 2000; Muller et al. 2000), especially in the face of increases in human demands and more variability in environmental conditions (Folke et al. 2004).

Adaptability

The adaptability of forests to changing external and internal factors is a key feature of ecosystems (Levin 2005). In many ways, the various structures, compositions, and functions of forest ecosystems develop in a way similar to that of individual species: they are constantly evolving in reaction to changes in their environment. Only through the ability to adapt have forest ecosystems been able to cover about one third of the global land area. Ecosystems do not evolve as a unitary whole, however. They are shaped, or self-organized, by interactions among individual components, which are changing in response to the environment, which itself is changing as a result of the interactions of the components and outside factors (Levin 2005). Selection of the best adapted species as well as the development of various functional relationships that are frequently non-linear and that cross hierarchical scales gives rise to complexity while also enhancing resilience by providing the ecosystem with the flexibility to respond to a wide variety of changes (Levin 2005; Drever et al. 2006; Solé and Bascompte 2006).

Forests are naturally always in a state of change, but the speed of change and the set of ecosystem components undergoing change are not constant. Larger natural disturbances and silvicultural treatments (i.e., managed disturbances) can play a special role in forests as they provide opportunities for drastic changes and adaptations to occur. Thus, the role of the various characteristics of ecosystem complexity becomes most evident after disturbances when forests go through a multifaceted reorganization phase (referred to as succession in forest ecology). Regardless of the type and severity of disturbances, it is far from certain that the new forest will be like the previous one, nor is that necessarily desirable. Quite the contrary, it is more likely that the future forest will be different in many, if not most aspects. Disturbances themselves are an inherent component of ecosystem development and therefore crucial for adaptation of forests to new altered conditions. They act at multiple spatial and temporal scales and favor species and interactions that are better suited to the new set of conditions, and thus are a crucial aspect in maintaining ecosystem function and processes.

Disturbance severity therefore interacts with a temporal component (forest succession) to determine the range of natural patterns of heterogeneity in forests. After very severe disturbances (e.g., volcanic explosions, glacial retreat), few or no components of the previous forests are available to act as a legacy and influence development of new forests. In these cases (referred to as primary succession), early successional stages likely have a limited number of species and interactions. Such forests have a simpler structural, compositional, and functional heterogeneity than forests in later successional stages or forests that were not severely disturbed and had components that acted as biological legacy elements. Spatial heterogeneity of structure, composition and function, and diversity of species tend to increase during times without disturbances. Thus, after a high-severity disturbance, forest ecosystems with sufficient time to develop can reach similar levels of diversity and spatial heterogeneity in structure, composition, and function as forests where disturbances are mainly of low severity.

Forest successional stages with low structural and compositional heterogeneity are widespread in natural forest landscapes and can provide for ecosystem diversity at the landscape scale. Where ecosystems and their component species have a long history of adaptation to a specific disturbance, heterogeneity in structure and processes is not a requirement for a resilient ecosystem. However, in the context of multiple, new, or unexpected disturbances, or even lack of the disturbance that has driven ecosystem development in the past (e.g., Attiwill 1994), the benefits of diversity in structure and composition for adaptability of forests to novel environmental conditions are well documented (Holling and Meffe 1996; Scherer-Lorentzen et al. 2005; Drever et al. 2006).

Silviculture and Complexity

Challenges for Silviculturists

Every forest fits the list of characteristics of a complex adaptive system. However, silviculture, as practiced on managed forests, has demonstrated a limited understanding of the implications of this very important aspect of forest ecosystems. As reviewed in chapter 2, silviculturists, inspired by the success in agriculture, have worked hard to reduce or eliminate many of the elements and behaviors inherent in complex ecosystem. The vision at the time was that a fully controlled and efficiently managed (i.e., homogenized) forest would best maximize the production of wood and other commodities. While heterogeneity of structure, composition, and function are not necessarily attributes of all complex adaptive systems (see example of Lodgepole pine), heavy-handed, top-down control, such as implemented in the most intense silvicultural practices, will greatly simplify and homogenize forests, and almost certainly will prevent the natural tendency of the system to readily adapt to new or recurrent disturbances or other environmental changes (Holling and Meffe 1996; Drever et al. 2006). Such an approach with a focus on order and predictability for each and every stand cannot be without consequences for future resilience and adaptability of forest ecosystems.

Moreover, fighting against the inherent behaviors of forest systems has proven to be challenging and silviculturists have learned that they often have to intervene heavily to maintain management objectives. Examples of such practices include the intensive vegetation control to minimize resource uptake by grasses, herbs, shrubs, or trees that are not of commercial interest. In fire-prone regions, thinning and fuel management practices aim to reduce the potential for fires that might destroy managed stands. Such controls are expensive, challenging, and often controversial, as evidenced by the debates about herbicide applications and fire management in many parts of the world. Silviculturists also dedicated much effort to reducing the direct impacts of disturbances, for example through sanitation cuttings.

The most profound challenge for silviculturists is therefore to accept that the reliance on "command and control" (Holling and Meffe 1996), which is at the heart of many current silvicultural practices, often works counter to one or more of the characteristics of complex adaptive systems. By embracing the notion that these characteristics are inherent and potentially desirable attributes of forested ecosystems and that they are influenced by silvicultural treatments, silviculturists will come to view their profession and their practice very differently.

In this light silvicultural treatments should be assessed in terms of their impact on each of the eight characteristics of complex adaptive systems. The reliance on the command-and-control model of top-down management, with its attendant belief in predictable outcomes, would thereby be purposely reduced. The focus of silviculture in managed forests would shift toward maintaining a full suite of possible outcomes so that the forest can readily adapt to new and modified conditions created by or following disturbances, be they from human or natural causes, or both. In doing so, silviculturists need to accept that some of the advantages and benefits of the traditional silviculture approach may be lost, and understand that "novel" benefits will be gained, many of which we may not currently anticipate. For example, regeneration practices will likely be more diverse, and the yield and quality of growth of individual trees and stands will likely be more variable. As a trade-off, forests will generally be more heterogeneous, more resilient, and better adapted to current and future biotic and abiotic conditions (Drever et al. 2006). Forests managed as complex adaptive systems are more likely to provide the increasing variety of services that humans expect from forests in the long term. This shift in management will likely result in fewer and less intense interventions and thus may prove to be less costly in the long run.

As we discussed in our review of the history of silviculture (chap. 1), the discipline changes in response to external factors or influences. Silviculturists cannot be expected to value complexity if society and landowners do not appreciate its significance as an intrinsic and important attribute of forest ecosystems. Appreciation comes from a better understanding of the importance of complexity to ecosystem functions and processes, such as adaptability to altered conditions, biodiversity, resilience, and productivity (Gunderson and Holling 2002; Scherer-Lorenzen et al. 2005; Drever et al. 2006). Certainly, our scientific understanding of these basic relationships is incomplete and discussions about specific theories and concepts are ongoing (e.g., see Chapin et al. 2000; Huston et al. 2000; Loreau et al. 2002; Naeem 2002; Tilman et al. 2002b; Hooper et al. 2005; Scherer-Lorenzen et al. 2007). However, it is generally agreed that there is a direct link between the maintenance of some heterogeneity and diversity of structures, compositions, and functions of ecosystems and the maintenance of its long-term productivity (Loreau et al. 2002; Scherer-Lorenzen et al. 2005).

The value of managing forests as complex adaptive systems will increase in light of expected future changes in social and environmental conditions. The potential benefits include a higher likelihood that forests are able to respond to a variety of changes. For example, the probability of exotic, invading plants, insects, and diseases is increasing with regional and global travel and trade. Trends such as altered resource levels or disturbance regimes due to projected climate change further strengthen the value of maintaining ecosystem resilience and adaptability (Woodwell and Mackenzie 1995; Folke et al. 2004). These issues will influence forests regardless of the landowner's management objectives, including intensive forest management with the goal of maximizing wood production at the lowest cost.

Maintaining the ability of forests to adapt to diverse and unexpected future disturbances without losing their ecological integrity should become a higher priority. Silviculturists cannot afford to wait until all aspects of complexity are agreed upon before considering the many potential benefits that such a new approach to managing forests could bring. We propose that the value of complexity science and thus of managing forests as complex adaptive systems is sufficiently well established that silviculture, as a discipline, will benefit greatly from adopting and adapting it.

Impacts of Traditional Silviculture on Structural Heterogeneity

The powerful effects of the top-down command-and-control approach to silviculture can be examined in more detail by contrasting the structural heterogeneity of three idealized forest systems: (1) intensively managed even-aged conifer plantations, (2) intensively managed uneven-aged stands, and (3) unmanaged temperate mixed-species forests (fig. 5.4). For

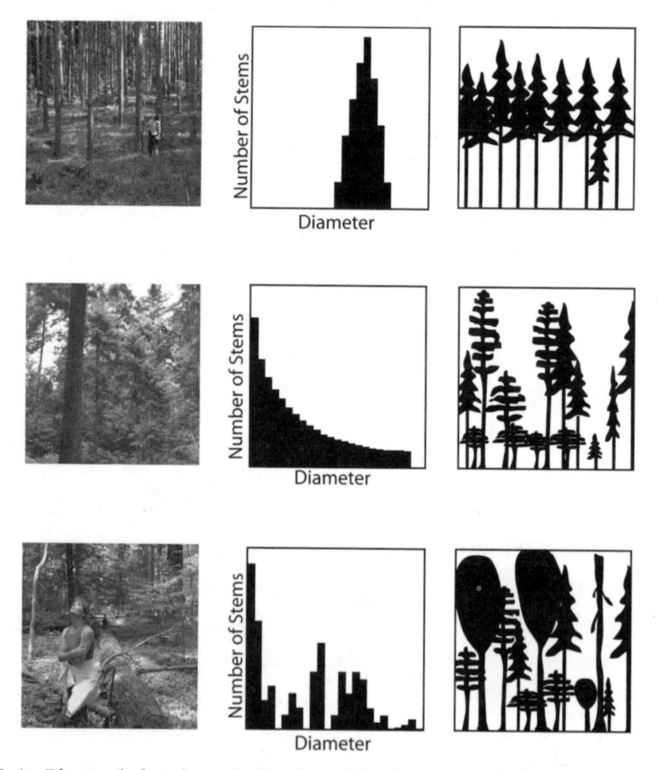

Figure 5.4. Photo (left column), diameter distribution (middle column), and conceptual drawing representing tree sizes of a mature single-species plantation in Scandinavia (top row), a single-tree selection forest in central Europe (middle row), and a natural mixed-wood forest in Québec, Canada (bottom row).

simplicity, we limit our argument to the discussion of the tree component of stand structure, but a similar comparison could be made for other ecosystem components. Our intent is to highlight differences among the three forests that are important indicators of potential resilience and adaptability of ecosystems. The managed even- and uneven-aged stands are less able to modify their structure and function in response to the external and internal factors affecting these forests. They are maintained within a narrow range of structural and compositional states by intensive interventions (fig. 5.5, shaded areas A and B). The unmanaged forest, in contrast, is responsive to changes and can develop toward a wide variety of states (fig. 5.5, shaded area C).

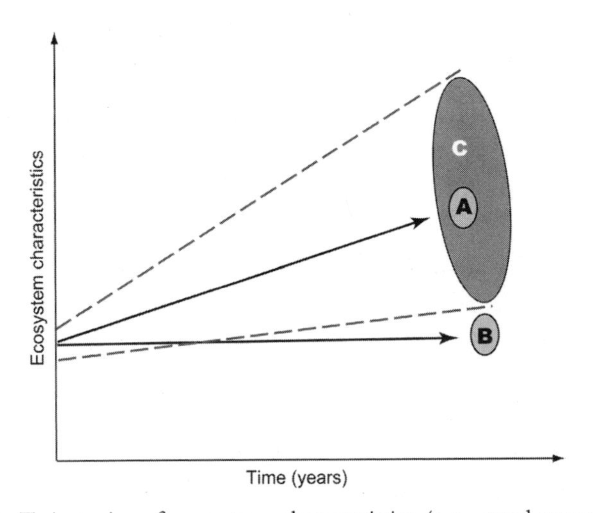

Figure 5.5. Trajectories of ecosystem characteristics (e.g., stand structure, timber volume, tree biomass, or habitat characteristics) for selected management scenarios. All forests are assumed to start from the same condition. The trajectory to point A represents a forest managed through single-tree selection, while the trajectory to point B represents the same forest managed as an intensive even-aged and single-species plantation. In both cases, ecosystem characteristics develop in a narrow predictable manner due to continuous and intensive management inputs. The shaded ellipse represented by C is much wider and characterizes the variety of possible developmental trajectories of forests either left alone or managed as complex adaptive systems. The variety of possible outcomes within C allows forests to be "creative" in adapting to new altered conditions.

Intensively managed even-aged conifer plantations, such as scotch, loblolly, and Monterey pines in Scandinavia, the southeastern United States, and New Zealand, respectively, are characterized by dominance of a single tree species with homogeneous spacing, stem diameter, height, and canopy characteristics within stands. Site preparation, release, and thinning treatments are aimed at maximizing productivity by homogenizing microsites and neighborhood conditions. Landscapes dominated by intensively managed plantations are composed of stands that are fairly similar in stand structure and composition, but differ in age and thus tree size and spacing. Such stands require fairly intensive management to be maintained because they are outside the range of natural conditions found in surrounding forests (see fig. 5.5, shaded area B). This is considered necessary to concentrate the productive potential of the site into the managed trees. The objective of management is to "combat" the inherent tendencies of such stands to move toward more diverse structural, compositional, and functional forests (fig. 5.5, shaded area C). Because of their lack of spatial and temporal structural and compositional heterogeneity, these stands have low resilience and are more likely to be threatened by disturbances (Drever et al. 2006) or climate change (Folke et al. 2004; Woods et al. 2005).

Intensively managed uneven-aged single-tree selection forests, such as Plenterwald in Switzerland, differ from intensively managed even-aged plantations in several important ways. They are made up of multiple (commonly two to four) over- and under-story tree species mixed at small spatial scales (the neighborhood scale, see chap. 4) and exhibit variability in tree species, size, canopy condition, and spacing within individual stands. Despite having a higher variability of structures than intensively managed even-aged plantations, managed uneven-aged stands are still being tightly regulated. Structural and compositional characteristics are relatively uniform both within and among stands. Although often described as "heterogeneous" or "complex" in the forestry literature, due to the tight control of species composition and diameter distribution, intensive uneven-aged management maintains stands within a narrow range of structures (fig. 5.5, shaded area A). Also, uneven-aged management approaches replicate similar structural and compositional patterns over the large landscape. Although the tree species composition and structural characteristics of this ecosystem can be within the range of conditions found in natural forests (fig. 5.5, shaded area C), the narrow variability at the landscape level reduces the range of conditions that would naturally occur without management. Consequently, the homogeneity of structures at both the stand and landscape levels is of concern when assessing ecosystem resilience due to its exceptionally high connectivity that can favor the spread of large-scale disturbance agents such as fire or insect epidemics (Andren 1994; Folke et al. 2004; Drever et al. 2006). The lower heterogeneity of structural and compositional conditions found both within and among stands likely reduces biodiversity. Furthermore, these forests require constant monitoring and intervention to ensure that they reach their narrow desired states.

Finally, unmanaged temperate mixed-species forests (*sensu* Peterken 1996), such as the hardwood forests in many regions in eastern Canada, have multiple tree species (often more than twenty), heterogeneous structure and composition, rich understory herbs and shrubs, and lots of dead material both standing and on the forest floor (Angers et al. 2005). Tree spacing, size, canopy conditions, and understory species composition (Crow et al. 2002; Angers et al. 2005) are typically quite variable at the neighborhood scale and from stand to stand. At all scales, heterogeneity of structures and processes occur and these forests change continuously over time with and without disturbance, maintaining what Gunderson and Holling (2002) called their "creativity" (fig. 5.5, anywhere in shaded area C). Other components of stand structure, such as understory vegetation, snags, and downed wood and their interactions, are all integral and influential components of the forests. Complexity processes (self-organization) are not impeded here, and they arise from the full suite of possible relationships that can develop among the various functional components of the forest. In this context, it is important to note that at any particular moment in time these forests can exhibit limited heterogeneity in structure and processes at certain scales, but this is likely to change again over time. Thus, developmental phases or systems with limited structural or compositional heterogeneity, such as during the stem exclusion phases after large-scale fires (e.g., lodgepole pine in Yellowstone Park or central British Columbia) or the later stages of boreal forest succession, still retain the essential characteristics of a dynamic complex adaptive system (Gauthier et al. 2000; Boucher et al. 2003).

To illustrate the impacts of a strict regulation approach and the successes of increased management efficiency through reduction of natural stand heterogeneity, we compare how typical diameter distributions vary among the three forest ecosystems described above over a range of spatial scales. While this is clearly a simplified characterization of the full heterogeneity found in forest ecosystems, utilizing a simple and common silvicultural descriptor makes the contrast more evident (fig. 5.6). The solid line in figure 5.6 shows diameter distribution considered typical of plantations at different stand ages. The dashed line in figure 5.6 represents a so-called J-shaped curve and is considered indicative of balanced uneven-aged stands (Smith et al. 1997; Schuetz 2001). The impact of the command-and-control approach on heterogeneity in the two managed forest ecosystems becomes obvious when the diameter distributions are related to spatial scales. Probability theory tells us that the diameter distribution of trees in a 100-hectare perfectly managed even-aged plantation or uneven-aged stand will change very little when samples are taken from a 0.25-, 1-, 10-, or 100-hectare plot. The distribution would also look very similar regardless of the specific location in which plots are located. An unbiased (but admittedly imprecise) estimate of the average stand diameter can be derived by measuring a single tree in an even-aged plantation.

In the uneven-aged forest the minimum sampling area needs to be large enough to include a basic tree neighborhood area, that is, trees from a range of diameter classes. Any increase in plot size will only improve the precision of the estimate. For plantations of similar ages and for uneven-aged stands, the stem diameter distribution should also not vary appreciably among stands and over time, assuming similar site quality and a constant management effort. As even-aged stands mature, the distribution will retain its basic shape (ignoring influences of thinning practices and asymmetric competition on distribution skewness) and shift toward larger sizes over time (fig. 5.6). The relative constancy of diameter distributions of even-aged plantations and uneven-aged stands in relation to sample plot location, plot size, and over time shows that both management approaches provide very little spatial variability at larger than tree

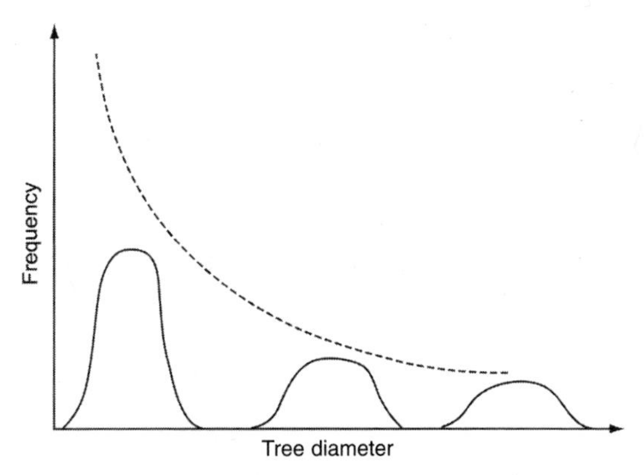

Figure 5.6. Typical diameter distribution of three even-aged stands (solid lines) at young (left), middle (middle), and older (right) stand ages. Diameter distribution of an uneven-aged stand is presented as a dashed line. Typical diameter distributions of natural stands don't exist and are therefore not presented here.

and neighborhood scales, respectively. In contrast, typical diameter distributions cannot be presented in graph form for the unmanaged natural hardwood forest, because no typical distribution exists. The shape and location of the diameter distribution will vary across a wide range of spatial scales and from one stand to another. Likely the distribution will vary in different portions of the stand and with plot sizes, and neither of these relationships will be consistent across stands.

Managing Complexity

To this point, chapter 5 has raised three main arguments. First, that all forests—even single-species plantations—behave like complex adaptive systems if left unmanaged. Second, that complexity is a highly desirable characteristic to maintain the adaptability of forests to a wide range of conditions that allow them to provide benefits for future generations. Third, that silvicultural approaches based on the agricultural model of top-down control limit the ability to manage forests as complex adaptive systems. We now need to address how complexity can be incorporated into silvicultural management and research in the context of the six attributes of complex systems outlined in chapter 3: (1) nonlinear relationships and not totally deterministic, quasi-chaotic behavior that makes predictions about the future dynamic of the forest uncertain; (2) boundaries and elements that are difficult to determine so that system limits are ill-defined; (3) openness to outside influences so that the system is never totally at equilibrium; (4) relationships among parts and processes of the system containing feedback loops that may cross scales or hierarchies of organization, making the system self-regulated or self-organized; (5) emergent behaviors that arise from interactions among parts and processes of the system that cannot be predicted from understanding of lower levels of organization; and (6) memory, such that previous states partially influence the present state of the system. Specifically, we focus our discussion on how silvicultural practices influence all these characteristics in the context of resiliency and adaptability of forest ecosystems.

Uncertain Future Conditions

Forests exhibit elements of quasi-chaotic and uncertain behaviors as a result of interactions among many nonlinear relationships. Silviculturists have expended a lot of effort fighting these behaviors to ensure a higher degree of predictability in forest development (chap. 2). Accepting unpredictability and heterogeneity as important and inherent characteristics of forests implies allowing forest development to follow a variety of possible paths. Development of forests includes many components of randomness (e.g., seed dispersal, herbivory, windstorms), but forests do not develop randomly. Instead, through well-understood successional processes, they develop toward what chaos scientists have labeled attractors (Solé and Bascompte 2006) or a range of conditions (as shown in fig. 5.5). The particular environmental conditions and interactions among many unmeasured factors will eventually and inevitably move the system toward the attractor. The attractors or range of conditions are not external properties to the system. They are produced by the variety of interacting elements within the system. By adapting complex systems theory, silviculturists should attempt to move to a prescribed envelope of possible future conditions for each stand, rather than forcing each stand to move to a specific condition, as described in yield tables or growth models (see fig. 5.5, shaded areas A and B). This envelope can be described as a distribution of possible stand conditions that reflects the attractors of natural forest succession (see fig. 5.5, shaded area C).

Forests Are Like Teenagers

There are many analogies between managing a forest as a complex adaptive system and raising teenagers to reach their full potential as adults. Like a forest, teenagers have the attributes of complex adaptive systems. Although parents endeavor to understand them, we cannot predict their future behavior, which makes some parents quite uncomfortable. To overcome this situation, some parents impose strict rules about behavior and dress, choice of friends, or career options. However, like complex adaptive forests where silviculturists have imposed strict management rules, teenagers will likely require ongoing strong interventions to achieve our narrow objective. By doing so, however, we increase the likelihood that they will be less able to deal by themselves with future challenges and surprises. Furthermore, they may rebel and exhibit extreme behaviors that will make us unhappy. In short, such "command and control" does not necessarily make teenagers more predictable or resilient. If we instead relax our grip and accept that we have given our teenagers the best upbringing possible, it is very likely that our son or daughter will develop into a responsible individual, although perhaps they will not become the person we had wished for or anticipated. But if the main attractor (i.e., our education, moral support, love, role modeling, and encouragement) is strong, the child we thought we knew so well is likely to surprise us (i.e., develop emergent properties) in many wonderful ways. They may actually develop into someone beyond our wildest dreams because their upbringing has prepared them to find their own specific solutions to challenges and surprises.

Giving up specificity in prediction of any particular future stand condition may seem like a step back in our efforts to manage forests on a scientific basis. Many forest owners require fairly exact predictions of future condition for each stand in order to assess the value of the ownership or calculate how much timber can be extracted annually in a sustainable way (and think that science can provide the tools to obtain this information). As we have argued, the science of complexity and assessments of impacts of disturbances on harvesting levels have shown that this is simply not possible. Silvicultural practices should ensure that forests do not develop outside the envelope defined by attractors of natural forest succession (see fig. 5.5, shaded area C) or that the envelope does not become too narrow. Descriptors of the distribution of possible outcomes, which could include mean conditions and associated variability around the mean, can be used to calculate management outputs. Thus, the idea of predictability is not abandoned but used at a level that is more appropriate for complex ecosystems. In this case predictability switches from single stands to a distribution of stand structures and compositions in the landscape, similar to the approach suggested by Hummel and Barbour (2007). Certainly, this approach requires further research on how to prepare such distributions for a variety of landscapes and ownership objectives and on how silviculturists can develop these distributions on their own; but, by loosening the grip on predictability, silviculturists may actually gain a lot of flexibility and save time and effort to combat the natural force acting in each and every stand.

There is an increasing variety of new modeling tools that can be used to simulate stands and landscape in more "complex" ways. Many are even able to incorporate changing conditions (i.e., new attractors) to predict the likely future conditions of the forest. Traditional growth and yield models that are fully deterministic and non-spatial (e.g., Daniels and Burkhart 1975) are not very useful in this context. More recent models that use trees as individual modeling agents and are spatially explicit (e.g., PTAEDA2: Burkhart et al. 2001) represent a significant advancement. Models that simulate forests by incorporating regeneration and growth routines at various spatial scales (e.g., SORTIE-ND: Coates et al. 2003) or even include stochastic elements (e.g., LANDIS-II: Mladenoff 2004) are even better suited to help silviculturists understand the envelope of desirable future stand structures.

Efforts of various research groups around the globe to develop stochastic and spatially explicit models of forest developments are encouraging. Recent developments in Bayesian Networks (for examples of

---

Bayesian Network: Models use a probabilistic, rather than a deterministic, approach to describe the relationships among variables. This approach to characterizing knowledge allows "driving" variables to be entered as a distribution of likely values (independent probability distributions). Outcomes are likewise expressed as probability distributions.

---

their use in ocean and fishery research, see Ver Hoef 1996; Lee and Rieman 1997; Borsuk et al. 2004) and linkages to climate change models will further improve forest simulation models. However, rather than approaching these models with a mind-set of improving predictability of forest development under specific conditions, development and use of the models should draw upon a solid understanding of the characteristics of complex adaptive systems, especially accepting uncertainty, unpredictability, and quasi-chaotic behaviors as intrinsic and desirable characteristics of any individual forest stand.

Since ecosystems are fundamentally a network of interacting elements, new models and modeling approaches need to be able to represent the important elements of the system both spatially and temporally. Complexity models require an organizational hierarchy to represent their system of interest (Parrott and Rok 2000; Green et al. 2005; Proulx 2007). To simulate the intricate functions of a forest, a model will need to represent, in a spatially explicit manner, the most important objects and functions that affect its short- and long-term dynamics at more than one spatial scale. Many hierarchical representations are possible, but in most cases they will encompass some or all of the following levels: individuals, which are lesser than populations, which are lesser than communities, which are lesser than ecosystems, which are lesser than biomes. Complexity theory also implies that it is not possible to simulate complex behaviors in stands by using whole stands as modeling agents because no interacting elements are present that can generate emergent properties at the stand scale. In fact, ecosystems structures, functions, and processes are now interpreted as emerging from inter-hierarchical interactions. For example, the slow (e.g., tree succession) and fast (e.g., insect dynamics) variables of Gunderson and Holling (2002) represent interactions occurring across two time scales. Complex behavior is always represented using a "bottom-up" approach to modeling. In such an approach, each hierarchical element is modeled as a discrete agent or object state, where each entity has functions that are characterized by relationships described by rules (or equations) and constant values or variables.

Modelers have used three general approaches to simulate ecosystem development: individual-based models, agent-based models, and cellular automata (Parrott 2002). Here, we present only the first approach and use SORTIE-ND (www.sortie-nd.org) as an example. In the SORTIE-ND model, the forest is represented by a large collection of interacting trees that are followed in both time (in steps of at least one year) and space. Those trees are currently divided among seedlings, saplings, and adult trees. Population-level dynamics are simulated by summing the collective activities of numerous individuals. Each tree is a discrete object that is described with various attributes (size, growth rate, age, crown morphology, and so on). Each tree's (individual) behavior is modeled with rules that describe the interactions with other individuals (e.g., effect of species and distance of neighbors on growth of individual trees) or its environment (e.g., growth of seedlings in relation to available light levels). Many of the interactions have nonlinear relationships and/or have random events associated with them. The nonlinearity of many interactions, the stochastic behavior of some objects and processes, and the large number of objects, rules, and stochastic events make SORTIE-ND a good example of a modeling approach aimed at being able to represent complex behavior in forests.

This is just one example of various models that can be used to simulate complex behavior at the forest stand scale. For any such model, the hierarchical levels being represented, the spatial and temporal scales used, and the functions and variables represented depend on the questions being asked, the available data, and the skill and approach used by the modelers. What is important to remember here are the basic elements that are required in such models to be used to simulate complex behaviors: (1) representation of many hierarchical levels, (2) representation of both spatial and temporal scales, (3) some stochasticity, (4) some nonlinearity, and (5) some representation of discrete entities or elements.

Ill-Defined Boundaries

If forests are viewed as complex adaptive systems with hard-to-determine boundaries, elements, and hierarchies, then all attributes or ecological processes occurring naturally within and around any individual stand are potentially important in maintaining its normal functioning or resilience, even those that may not seem important to us (McCann 2000). Outside influences are therefore an inherent characteristic of forest ecosystem dynamics and should be managed as such. These influences act in a variety of dimensions, including ecological, economic, and social. While harvest scheduling and regulations acknowledge that stands are not isolated in the landscape, management has typically not taken into account the full set of implications of the juxtaposition of forests. In most cases where spatial context or adjacency has been considered, it was done because of legal constraints (e.g., limits to clearcut size or "green-up" constraints; see Brumelle et al. 1998) or to reduce tree damage, such as wind protection. All silvicultural practices should be evaluated not only for their impacts on the treatment area, but also for their impacts in adjacent forests, agricultural areas, or urban landscapes. This assessment should not be limited to landscapes, but should be applied to many ranges of spatial and temporal scales that match the range of structures, processes, and functions that are manipulated.

The emphasis on stand-level management needs to be reduced or eliminated when outside influences are an important characteristic of a system. Species mobility (fig. 5.3) provides an example of a hard-to-determine boundary. It is a characteristic that would not necessarily be considered in a thinning prescription that focused on timber production, especially if the species was not present in the stand. An opportunity for an animal to move across the stand may be critical to the long-term survival of the species. Thus, outside influences may suggest altering thinning prescriptions. Other issues related to wildlife habitat at multiple scales (both larger and smaller than stand scales) provide similar examples (Wilson and Puettmann 2007). Management practices need to be developed, applied, and assessed at multiple scales and in multiple dimensions, such as how much the expected ecosystem development deviates from natural trends, impacts on various functions and processes, and impacts on structural heterogeneity at the neighborhood, stand, and landscape scales. To promote ecosystem adaptability to a wide variety of disturbances, variability should exist at multiple scales, starting from local tree neighborhoods to stands, landscapes, regions, or ownerships.

Outside disturbances need to be viewed as an inherent part of forest ecosystems. The role of current growth models provides an example of the impact of this change in view. Yield tables and growth models are perceived as reliable predictors of tree and stand growth even though their predictions are generally valid only in the absence of disturbance. Their reputation as reliable is contingent on the view that disturbances are external factors. If disturbances are accepted as an inherent part of forest development, current growth and yield predictions would be understood to be inaccurate. In most cases they overpredict yields because disturbances typically result in reduced stand growth. Thus, the acceptance of disturbances as an integral part of ecosystems equates to accepting unpredictability of forest development and puts the apparent power of current growth models into a different light. It requires a humble acknowledgment that silviculturists do not have a solid understanding of all factors that influence forest development and growth and suggests the need for developing models better able to incorporate uncertainty.

Never at Equilibrium

One important characteristic of ecosystems is that they are never at equilibrium (Levin 2005). Changes in the system provide constant feedback to the system. This feedback allows systems to adapt to the ever-changing biotic and abiotic conditions. Managing forests as complex adaptive systems means accepting the view that ecosystem structures and processes are continuously changing and this change is an important characteristic that helps ecosystems respond to environmental change. Silviculturists have historically managed forests to maintain a narrow set of characteristics. These characteristics may include a limited set of species, regular spacing, and uniform tree and crown sizes in even-aged stands, or a diameter distribution that changes little over space or time in uneven-aged stands (see fig. 5.4). The notion of being able to achieve stability and constancy is an inherent feature of the command-and-control approach. Instead, silvicultural practices should be assessed in terms of their impacts on the variety of dynamic properties found in forests with the understanding that these dynamics act over multiple spatial and temporal scales. In general, practices that do not stifle, but rather accommodate, dynamic behaviors are likely to facilitate resilience and adaptation in forest ecosystems.

Self-Regulated

Self-regulation in complex systems occurs mainly through positive and negative feedback loops. Addressing feedback loops in management strategies presents a major challenge because little is known about relationships that cross hierarchical scales, at least not in a context that provides operational solutions for silviculturists. Learning about such relationships will require new, multiscaled research approaches because the answers will not be found in a single field study (Ganio and Puettmann 2008). For example, site preparation or vegetation control quickly alters shrub, herb, and grass communities which, in turn, affect the overall dynamics of the forest. The response of insect or small mammal communities that eat the seeds from these understory plants may be delayed. Predator populations that keep insect and small mammal populations in check through negative feedback processes may be influenced at even longer time scales and much larger spatial scales. Thus it may be very difficult for either a researcher or a silviculturist to draw a direct link between a popular site preparation practice that efficiently reduces vegetative competition and increases in animal damage to plantations that may occur years to decades later.

Another relevant example is the effect that vegetation control or other silvicultural practices may have on the invasion rates of exotic plant species that may, in turn, modify fire regimes, leading to long-term shifts in ecosystem functions, processes, resiliency, and adaptability (D'Antonio et al. 2000).

Feedback loops that cross hierarchical scales are one more reason for silviculturists to ensure that all ecosystem components are managed and maintained at functional levels. As many of these cascading interactions are not sufficiently understood, a precautionary approach similar to the "coarse filter approach" to wildlife management (Seymour and Hunter 1999) would be a useful starting point in order to maintain all potentially important elements and processes until more information is available, especially since the impacts of exotic invaders and climate change are expected to increase in the future (Folke et al. 2004; Steffen et al. 2004).

Develop Unexpected Properties

Emergent properties (as defined earlier, p. 119) are unexpected phenomena that result from interactions among individual components of forests. The spontaneity and unpredictability of emergent properties are viewed as an important factor in ecosystem resilience. This "creativity," in a sense similar to genetic recombination and mutation, provides opportunities for forests to adapt to new conditions. Attempts to model this creativity highlight the challenge (or impossibility) of providing simple silvicultural guidelines to manage for emergent properties. By definition, simulation models that are based on linear relationships, purely deterministic behaviors, and closed system assumptions cannot develop emergent properties. The development of operational forest models that can handle basic principles related to emergent properties, such as nonlinear feedback loops, is still in its infancy. Early attempts are limited to a few selected ecosystem processes and structures (e.g., Breckling et al. 2005). Development of operational models that allow investigation of patterns that lead to emergent properties will be especially important for assessing the impacts of projected climate change on forests. Climate change will undoubtedly impact forests in unpredictable ways that may be negative or positive for forest management. The silviculture of the future may well be focused on trying to reduce negative emergent properties from a societal needs-and-necessities perspective. Maintaining the full suite of characteristics of complex adaptive systems in managed forests may provide the highest likelihood of desirable emergent properties (Folke et al. 2004).

Affected by Initial Conditions or Previous States

The sixth attribute of complex adaptive systems is that they remember previous states, which can have a great influence on current conditions and future developments in forests. This characteristic of a complex system is probably one of the easiest to understand for silviculturists. For example, early management efforts, such as coppice systems, took advantage of this memory to encourage hardwood regeneration. Later practices aimed at eliminating the memory of previous states, the most common being the removal of hardwood sprouts or shrub vegetation as part of vegetation control efforts and the removal of advanced or natural regeneration in tree plantations. Present-day structural retention, or management of legacies in many regions by silviculturists, aimed at retaining live green trees, snags, or downed wood as habitat structures (e.g. Franklin et al. 1997; Mehrani-Mylany and Hauk 2004), is an example of memory management in forests. Embracing memory as an inherent feature of forests requires that the concept of legacies be expanded beyond trees. All structural components, including herbaceous layers (Roberts 2004) and shrub or mycorrhizal communities, are to be viewed as legacies, as minor differences in initial conditions can have great impacts on the development of complex systems (Solé and Bascompte 2006). The legacy concept in silviculture needs to be expanded to cover more than just the harvesting operation and retention of structure. For example, in even-aged stands, the open habitat conditions found during early stand development could be viewed as legacies when stands mature. The legacies concept in complex adaptive systems also goes beyond structural management. Silviculturists should think of processes and functions such as organic matter decompositions (e.g., Høiland and Bendiksen 1996; Nordén and Paltto 2001) as legacies that provide memory to forest ecosystems.

Silviculturists are already managing for some of the system attributes listed above without necessarily having a complete conceptual understanding of how complexity develops in forests (fig. 5.7). Work by the Pro-Silva group in Europe (www.prosilvaeurope.org), for example, and

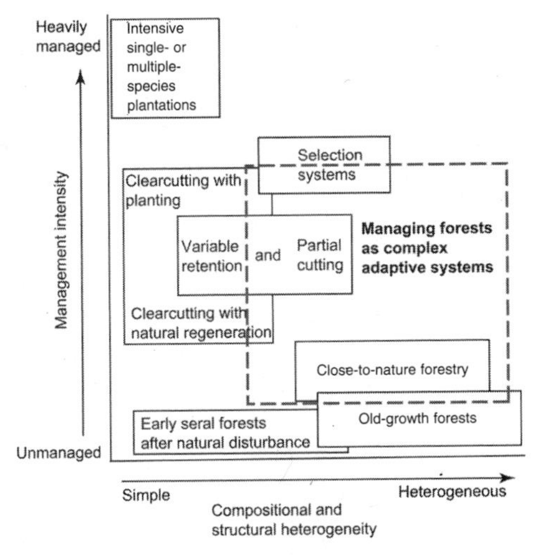

Figure 5.7. Selected silvicultural treatments aligned along gradients of heterogeneity (x-axis) and management intensity (y-axis). The proposed approach of "managing forests as complex adaptive systems" is represented by a box with dashed lines. The new approach covers a wide range of management intensities at a variety of spatial scales. The wide range of compositional and structural heterogeneity represents variability at different spatial scales, from within and among stands, that stems from the wide envelope of possible developments that forests can follow (see figure 5.5).

the various ecosystem management approaches in North America (e.g., Kohm and Franklin 1997; Bergeron et al. 2002) are clearly moving silviculture toward addressing some of the issues discussed in this book. Learning from complexity science and managing forests as complex adaptive systems can further the conceptual development of these trends. It will allow the discipline of silviculture to move toward an internally consistent scientific framework. The list of system attributes described in this chapter can be used to develop specific and localized reference measures to assess silvicultural practices. Thus, acknowledging and managing forests as complex adaptive systems is part of the continuous development of silviculture. It will help silviculturists to manage forests for resilience and adaptation in the face of changing environments and societal needs (Platt 1994; Holling and Meffe 1996; Drever et al. 2006).

Steps toward Managing Forests as Complex Adaptive Systems: Where to Start?

We have argued that developing a silviculture for managing complexity requires a shift in basic approaches to silviculture. Readers will rightfully ask how such shift would exhibit itself in the day-to-day activities of silviculturists. What follows is a list of practical ideas that can be incorporated into their activities right away in an attempt to move toward managing complexity in forests. In putting this list together, we took advantage of the strength of traditional silviculture and of numerous recent trends in silvicultural research and application. With this emphasis, we hope to demonstrate not only that managing forests as complex adaptive systems is useful as a guiding concept, but that many aspects of the approach have already received attention and may be at a stage where silviculturists can consider their implementation. We tried to cover a wide variety of bases/topics, and consequently readers will find some examples more applicable to their situations than others. The list is intended to be a catalyst that encourages silviculturists to assess their work for more opportunities to manage forests as complex adaptive systems. For a more general discussion about managing resiliency, see the Resiliency Alliance workbooks (http://wiki.resalliance.org).

Applying a Diversity of Silvicultural Treatments at Various Spatial and Temporal Scales

Every ecological process operates across a characteristic range of spatial and temporal scales, and this needs to be recognized and translated into management prescriptions. Silvicultural activities have traditionally been evaluated based on their impacts on the scale of a stand over a time period of forty to one hundred years. However, the many other processes that contribute to ecological phenomena such as soil hydrology, biodiversity, carbon cycling, and resilience act on very different spatial and temporal scales (fig. 5.8). Silviculturists need to separate the range of temporal and spatial scales of processes impacted by management prescriptions from the range of scales at which their management goals have traditionally been defined and assessed. A clearer separation of goals and impacts will help in addressing the challenge of developing new planning tools and techniques that can accommodate a variety of scales.

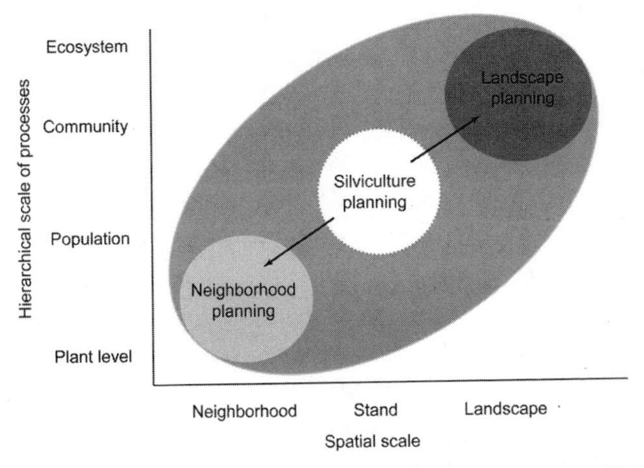

Figure 5.8. Processes and spatial scales related to forest management. Traditional silvicultural practices and books typically target the stand scale. Approaches such as those summarized under the label "ecosystem management" relate silvicultural issues from the stand to the landscape scale. Approaches that relate silvicultural practices to smaller scales—for example, neighborhood scales—have had limited influence on development of silvicultural approaches and practices despite many important variables and processes that influence stand-level dynamics acting at these smaller scales.

In managing forests as complex adaptive systems, management success can no longer be defined by single measures such as percent stocking or cubic meter per hectare per year (Loehle et al. 2002). Instead, success will be measured by a combination of spatial and temporal measures, ranging from micro-scale (e.g., proportion of stand that provides habitat for an endangered shrew), stand scale (e.g., timber production), and watershed scale (e.g., salmon habitat) to landscape scale (e.g., visual quality, carbon sequestration, animal migration, resilience, and even forest migration due to climate change effects) (Hann et al. 2001; Wilson and Puettmann 2007).

To begin with, treatments should specifically consider the scale of the processes that are managed and they should be applied at that scale. For example, vegetation control practices should acknowledge that seedlings typically interact with neighboring vegetation over short distances, no more than a few meters (e.g., Wagner and Radosevich 1991, 1998), so that there is no need to treat the whole stand to free one-meter tall seedlings from competition. Growth interactions among adult trees are another neighborhood phenomenon, found to generally act up to ten to twenty meters, but often much less (e.g., Canham et al. 2004; D'Amato and Puettmann 2004). Thinning prescriptions aimed at encouraging tree growth should also be implemented at this scale. Thus, prescriptions can accommodate variation of specific local conditions at their proper scale. This approach not only may provide benefits in terms of ecosystem diversity and adaptability, it may even lead to higher yields than prescriptions planned at the stand scale (Marshall et al. 1992).

Forests are influenced by dynamic, multifaceted disturbance regimes that include a wide range of disturbance patterns, agents, sizes, frequencies, and intensities (Frelich 2002). Reflecting this, silvicultural treatments should be varied within landscapes, ownerships, and even stands. Most tree species can be managed with multiple silvicultural systems, especially if full stocking by a single species is not required on all sites. Integrating information about scales of the various natural disturbances into management targets and goals should be reflected in silvicultural prescriptions.

Providing variability among stands can also be accommodated rather quickly. For example, rather than requiring 80 percent stocking of commercial tree species in all stands, programs could be changed to allow stocking of stands within a region or ownership to vary from 50 to 100 percent, but still reach an average of 80 percent. The actual range of values should be determined in cooperation with disturbance ecologists to match the natural range of variability. This approach would be inexpensive to implement, provide for a wider range of stand conditions, encourage natural regeneration and species mixtures, and reduce management efforts required to bring every single stand up to standard.

Spatial or temporal scales should not be viewed in isolation. Silvicultural activities should instead be based on hierarchical planning levels that consider ecosystem responses at plant, neighborhood, stand, and landscape scales over one, five, twenty, and one hundred years. Current management practices are starting points that will gain value through modifications that increase their abilities to accommodate multiple scales. For example, thinning practices can be modified by leaving gaps or uncut islands, by varying density in response to local stand or soil conditions, or by following specific guidelines for tree selections, such as maintaining trees from all parts of the diameter distributions or leaving minor species (Cissel et al. 2006; Wilson and Puettmann 2007). At the same time, thinning can accommodate aspects at the single-tree scale, such as leaving snags or releasing trees with unique crown structures that provide nesting opportunities.

Monitoring a Wider Variety of Descriptors and Moving beyond the Stand Concept

A more complete appreciation of the workings and behavior of forest ecosystems is required by silviculturists. The tree-focused stand descriptors (chap. 2), so standard in silviculture for so long, are not adequate to describe the full heterogeneity of species, structures, and processes managed by silviculturists. Plant species other than trees, insects, fungi, lichens, birds, and mammals all play important roles in functional ecosystems, and all are affected by silvicultural treatments. The characterization of the diversity within ecosystems is moving toward using "functional groups" of ecologically similar species (Kolb and Diekmann 2005; Aubin et al. 2007), which reduces the number of entities the silviculturist needs to consider. While tree-centered structural attributes are clearly important in forests (e.g., snags, large live trees, dead wood on the ground), they are often not sufficient to describe all ecosystem functions. Aubin et al. (2007) and Angers et al. (2005) use understory vegetation functional group diversity and structural variability, while Beaudet et al. (2004) and Bartemucci et al. (2006) use the vertical and horizontal variability of understory light as indicators.

A silviculture that thinks "beyond the trees" will be better suited to contribute to a wide variety of natural resource management issues. Accepting within-stand variability will actually free silviculture from some of the strings associated with timber production, such as high regeneration success on every single hectare of every single stand. It will allow silviculture to more effectively engage in a variety of settings that have previously provided special challenges, including forest restoration efforts (Frelich and Puettmann 1999; Sarr et al. 2004; Sarr and Puettmann 2008), biodiversity protection or enhancement (Angelstam 1998; Kuuluvainen 2002), and management for resilience (Bengtsson et al. 2003; Drever et al. 2006).

Managing forests as complex adaptive systems also requires a new definition of the stand concept, especially in terms of its relationship to spatial variability and heterogeneity. It implies valuing a wider variety of stand structural and compositional possibilities at multiple spatial scales and including them in inventory and planning documents. Silvicultural treatments should aim to maintain as much species, functional, and structural diversity as possible (e.g., Mason and Kerr 2004).

To address spatial variability and heterogeneity, silviculturists need to be able to assess and monitor structural and compositional heterogeneity and variability in inventories. This may imply changes in sampling design, including plot layout and/or locations. Sampling schemes employed in ecological studies are better suited to describe stand heterogeneity and variability than typical silvicultural plot samples and could be adapted for management purposes. For example, the line transect and line or point intercept methods (Thompson 2002) allow calculation of overall stand average as well as spatial variability and heterogeneity. Nevertheless, even current inventory data can provide some useful information if variability in data are reported and used. With the wide availability of computing software, it has become easier and cheaper to store, compute, and present information about both within- and among-stand heterogeneity and variability. Stand summaries should be expanded to report more than mean conditions and should include standard deviations, 95 percent confidence intervals, and spatial autocorrelation.

New technologies are available to help with the challenge of working at multiple spatial scales. For example, by utilizing tools such as GPS, GIS, satellite, and air photos (Couteron et al. 2005) or newer remote sensing technologies such as LIDAR (Lefsky et al. 2005), silviculturists can obtain better information about stand variability and heterogeneity. Computing technology, such as field data recorders with built-in GPS capabilities, will help to reduce the additional burden in planning activities by simplifying documentation of spatially variable silvicultural prescriptions.

Research on plant neighborhood effects on growth (e.g., Canham et al. 2004; D'Amato and Puettmann 2004; Stadt et al. 2007) and growth efficiency (Mainwaring and Maguire 2004) shows that the more heterogeneous the stand is at small scales, the more important it is to have spatially explicit information to accurately predict growth. Spatially explicit forest models are now quite useful for development and application of silvicultural treatments (Amateis et al. 1995; Pretzsch et al. 2002; Coates et al. 2003; Radtke et al. 2003). These models provide more flexibility to explore new and innovative silvicultural prescriptions to manage forests for greater variety of stand structures (Courbaud et al. 2001). As these models become more process-based (e.g., Miina and Pukkala 2002), they become more useful for investigating a full variety of ecosystem processes and functions that are increasingly becoming available to silviculturists in simulation models. These models may not need to be run for every stand that is scheduled to be treated, but when scientists and practicing silviculturists work together they can serve as guides to help silviculturists understand important principles and help predict possible ranges of stand development over time and space.

Spatial descriptors are needed to better characterize many important processes and drivers influencing ecosystem development, especially for the gap-scale disturbances that occur at a smaller scale than the stand scale (e.g., Brokaw 1985; Spies et al. 1990; Denslow and Hartshorn 1994;

Coates and Burton 1997; Kneeshaw and Bergeron 1998). Silvicultural treatments should re-create as much as possible the variability of natural disturbances, both within and among stands, so as to allow the forest to experiment with various solutions to changing conditions created by the treatment and continuously changing climate (Folke et al. 2004). To do so, silviculturists need to adjust their current understanding of how forest ecosystems work with predictions about future environmental conditions. This will help determine what type of natural disturbance to emulate and at what scale, which ecosystem characteristics are most influenced by silvicultural practices, and what treatments are most appropriate to implement as stands mature in order to maintain a functional forest ecosystem.

The vertical dimension is another critical component of the forest as a complex adaptive system. For example, the amount and vertical distribution of leaf area or canopy layers have been linked to tree and stand growth in various forest ecosystems (e.g., O'Hara 1989; Smith and Long 1989; Seymour and Kenefic 2002; Dean 2004; O'Hara and Nagel 2006) and wildlife niche differentiation and food web complexity (e.g., MacArthur et al. 1962). While the importance of canopy layering on productivity in mixed-species stands has been known for some time (Assmann 1961), the recent focus on canopy structure and leaf area distributions has led to a better understanding of the impact of different silvicultural practices on vertical structure and thus tree and stand growth. For example, an appreciation of the importance of the range of tree sizes may help understand potential thinning responses in even-aged stands (O'Hara 1989). Management of canopy depth may be a suitable tool for managing carbon allocation patterns through the relative allocation to branch- and stemwood produced by trees (Smith and Long 1989) and thus may influence carbon budgets. Similar approaches apply to belowground heterogeneity in rooting patterns (Rothe and Binkley 2001).

In multiaged forests, the variability in canopy conditions is higher yet, and the influence of canopy structure on growth efficiency of trees has been demonstrated in a variety of forest ecosystems (e.g., Seymour and Kenefic 2002; O'Hara and Nagel 2006). While data about leaf area distributions may not be available for typical stands, future research can provide information about the impacts of different vertical distributions, and efforts are needed to translate these patterns into parameters that silviculturists can use in planning and implementing practices.

Incorporating Risk and Uncertainty into Management

To manage forests as complex adaptive systems it is important to acknowledge and incorporate risk and uncertainty into everyday forestry practices (Backéus et al. 2005; Ericksson 2006). As more knowledge about short- and long-term implications of disturbances becomes available, it needs to be utilized in silvicultural applications and predictions (Thorsen and Helles 1998). By their nature, it is impossible to plan for specific stochastic events. Instead, silviculturists should view disturbances and associated impacts on ecosystems in a similar manner to an insurance company. Insurance companies do not calculate whether or not a specific house will burn. Instead, they have developed a very successful business model by utilizing information about fire probabilities to calculate insurance premiums for specific houses. Accepting a wider range of possible outcomes for individual stands (see fig. 5.5) will require acceptance of variability within and among stands. This may mean, for example, that moderate seedling mortality will not automatically result in replanting efforts, especially when regeneration of neighboring stands has been quite successful. Accepting stochastic elements as an inherent part of ecosystems is also important for management of expectations (Rivington et al. 2007; see also earlier discussion about prediction models). In this context, fully stocked, undisturbed forests are not viewed anymore as the norm. Consequently, deviations from fully stocked stands, for example due to windstorms, snow breakage, or insect problems, are not automatically interpreted as management catastrophes and should not reflect negatively on job evaluations or reputations of managing silviculturists (unless obvious mistakes have been made). Thus, incorporating risk and uncertainty does not necessarily have to result in an adjustment of tried-and-true management approaches (for example, see González et al. 2006). Instead it should be interpreted as an opportunity to avoid having to impose a narrow range of stand structures on every single stand (Hummel and Barbour 2007). It provides flexibility for silviculturists to use a wider variety of treatments and to carefully weigh responses to unplanned events and disturbances, including simply accepting them as an inherent and therefore valuable part of complex, adaptive ecosystems.

Developing Gradient- and Process-Based Silvicultural Research

The shift toward viewing forests as complex adaptive systems also requires a different approach to silvicultural research and education (see chaps. 2 and 4). Three avenues of research are of special interest, including (1) specifically defining the scope of inference, (2) providing information about trade-offs and gradients for treatment choices, and (3) increasing the generality of findings by focusing studies on underlying ecological principles.

First, we suggest that researchers specifically address the scope of inference and how scaling of research results to management applications influences practical applications (e.g., Cissel et al. 2006). In scaling up, questions about the variability in study results and how this could express itself in operational settings are of special interest. It is important to consider that the scope of inference is limited not to spatial components, but to other dimensions such as climatic, economic, and social conditions.

Second, we suggest that silvicultural research provides information that allows managers to assess trade-offs among choices along a wide gradient of treatments. For example, thinning studies often compare a limited set of replicated densities. Instead, with similar research efforts, studies that present information about changes along density gradients provide much more flexibility for silvicultural applications.

Third, silviculture research should strive for generality and theories and investigate basic response patterns or "conceptual generalizations based on the understanding of the involved processes" (Zeide 2001b, 49; see also chap. 2). Because of their manipulative nature, many silvicultural studies are better suited than observational studies often used by ecologists to investigate basic mechanisms of ecosystem responses to treatments or disturbances. Equally important, when research studies investigate how treatments affect processes that underlie growth patterns, information will become of general interest (*sensu* Zeide 2001a, b). Information about important processes and drivers of ecosystem development may have a wider inference scope and may be more broadly applicable in a wide variety of environmental conditions than information about the amount of response (Zeide 2001b). It may even provide information for new conditions, such as those expected with climate change. For example, additional measurements of resource levels or microclimate conditions greatly improve the understanding of vegetation control or thinning applications. These measurements will help sort out the relative importance of moisture, light, or other factors. While models that describe processes or functions may not be directly translatable into prescriptions, they are valuable as references for calibration and validation of new tools (e.g., O'Hara et al. 2001). Studying basic ecological relationships sends the message to practitioners that research provides information for managers to develop silvicultural treatments, rather than suggesting best treatments. Finally, it will go a long way in improving the reputation of silviculture as a scientific discipline that deserves a prominent place in research programs and institutions.

Conclusion

We conclude by summarizing four key principles that need to be incorporated into silviculture to accommodate forests as complex adaptive systems:

Consider as wide a variety of ecosystem components (i.e., more than trees) and functions as possible. The list of characteristics used to describe complex adaptive systems (see chap. 3) provides a basis for developing specific assessment criteria.

Abandon the command-and-control approach. Management of forests should accept variability in space and time as an inherent attribute that allows forests to adapt to new internal and external biotic and abiotic conditions.

Actively maintain and develop within- and among-stand heterogeneity in ecosystem structure, composition, and function to recreate the natural variability in forest conditions and processes.

Allow stands to develop within an envelope of possible conditions. Predict and measure success at the landscape scale rather than the stand scale and allow for multiple development trajectories at lower scales.

Managing forests as complex adaptive systems has several implications or outcomes. First, forests that have developed heterogeneous structure, function, and composition rather than being managed to a specific, narrow set of stand structures tend to be better able to adapt to changing abiotic and biotic conditions. This adaptability is especially critical because of the rapid pace of climate change and species invasions.

A second advantage of managing forests as complex adaptive systems is the reduced need for "command and control" (*sensu* Holling and Meffe 1996). Accepting unpredictability as an inherent feature of forests decreases the emphasis on managing all forests according to a single set of "best" management practices. It therefore requires less vigilance from silviculturists who can accept a range of developments as long as the whole forest achieves economic, social, and ecological objectives. In many cases, this will result in lower costs, reduced ecological impacts, and higher social acceptance.

Silvicultural research and prescriptions should be based on the knowledge that nonlinear, interrelated causes and feedback loops that span hierarchical levels of organization and encompass many spatial and temporal scales are all inherent features of ecosystems. It is this multiplicity of factors occurring at various scales that is necessary to allow forests to recover quickly after a wide range of disturbances, adapt to climate change, and resist species invasions.

Modifying management practices to accommodate ecosystem resilience and adaptability requires appropriate changes in research and educational approaches. Much can be learned from interactions with ecologists and other complexity scientists, as they have focused on understanding the complexity of ecosystems for a long time. With an open and critical mind, silvicultural researchers and educators can learn to appreciate different perspectives, use new tools and techniques, and thus contribute to an improved understanding and management of productive, resilient, and adaptable ecosystems.

The proposed changes pose deep philosophical and practical challenges to current silvicultural thinking. It will not be easy for many silviculturists to abandon the agricultural view that forests are controllable systems and that efficient management requires homogenization of stand structures. Awareness of and openness to the work of complexity scientists will help silviculture make this shift. Furthermore, complexity science provides a conceptual framework for many of the modifications and adaptations to silvicultural prescriptions that have already been implemented in the recent decades. It is a valuable template that can guide the further development of new silvicultural approaches and practices.

Glossary

afforestation. Establishment of trees in areas that did not support forests before

age class. Group of trees about the same age

age-class distribution. Distribution (by area) of stand ages. Balanced age-class distributions have the same amount of area in each age class.

agroforestry. A land-use system that combines agriculture and forestry on the same ground

agronomy. The science of growing plants for food, fuel, and fiber

analysis of variance. Group of statistical models that assigns portion of variances to explanatory variables

annual allowable cut. Amount of wood allowed to be harvested in one year, calculated to ensure timber sustainability

artificial pruning. Removal of branches to improve timber quality

autecology. Science investigating the interaction of species and their environment

basal area. Cross-sectional area of a tree at breast height, often summed up to an area basis (usually per hectare)

biodiversity. The diversity of taxa and biological processes found at all levels of the ecological hierarchy (genes, species, community, ecosystem)

biological legacy. Life form, propagule, organic structure, or material (or its footprint) remaining after a disturbance

broadleaves. Trees having relatively broad rather than needle-like leaves. Typical broadleaf trees are maple, birch, and oak species.

chaos theory. Apparently random behavior of nonlinear, dynamic systems

clearcut. Area in which all or most trees have been harvested recently

climax state. A later stage of succession, in which plant communities are fairly stable

community computer model. A computer program, using mathematical equations, that attempts to simulate the dynamic of a group of organisms

competition. Negative interactions among species for resources such as nutrients, food, water, territory, and light. Competition is believed to play an important role in structuring communities and is one of the main forces behind natural selection.

competitive exclusion. Theory that predicts that two species with the same requirements can't coexist

complexity. A concept that characterizes something or a system (such as a forest) with many interrelated parts that are interacting among each other

continuous cover (*Dauerwald*). Management form that requires continuous cover of trees at all time

control (or check) method. Inventory method in which 100 percent of trees are measured and inventory data are used to determine harvest operations, usually associated with uneven-aged management

coppice. Management form that relies on vegetative reproduction ability of trees (usually from the stump or roots) for regeneration. Typically used for firewood production.

coppice with standard. Coppiced stands in which selected trees are maintained through multiple cutting cycles

crop tree. Trees that silviculturists favor because of their desirable attributes

cutting cycle. Time between two continuous harvests

designer ecosystems. Ecosystems modified to ensure that they continue to provide services in a human-dominated world

deterministic science. Approach to science that assumes any phenomenon can be predicted based on a chain of prior events

diameter distribution. A measure of the number of trees in different diameter classes within a stand.

disturbance. Any sudden, temporary, and relatively rare event that causes a profound change in the dynamic of a system. Typical natural ecological disturbances are fires, flooding, windstorm, and insect outbreak. A typical anthropogenic disturbance is clearcutting.

disturbance regime. Any recurrent disturbance that tends to occur in a forested area. It is often defined in terms of timing, frequency, predictability, and severity.

ecological restoration. Management to restore ecosystems that have been damaged by natural or human disturbances

ecology. The science that studies the interaction of plants and animals with their environment

economic liberalism. Economic theory that advocates minimal interference of government in the economy

ecosystem. A natural area or unit consisting of interacting living organisms controlled by the same physical factors of the environment

ecosystem-based management. An approach to natural resource management that takes into consideration the whole ecosystem functioning instead of focusing only on one particular attribute of the system such as tree production

ecosystem function. The almost infinite interactions and processes that characterize an ecosystem such as nutrient cycling, soil development, water filtering, and so on

ecosystem process. Any well-defined ecological dynamics, such as productivity, succession, or decomposition

environmental gradient. Gradual change of plant communities and environmental conditions

even-aged stand. Stand in which all trees are of similar age

evolution. In biology, the change in the inherited traits of a population from one generation to the next

facilitation. A successional process by which a species modifies its environment, which in turn facilitates the establishment and growth of another species. Shade-intolerant tree species are often thought of as facilitating the establishment and growth of shade-tolerant trees by providing some shade.

Feller buncher. Harvesting machine that cuts trees in place and then places them in bunches for transportation

fertilization. Nutrient addition to improve tree growth and health

firewood. Wood used for heating purposes, usually from small-diameter trees

fitness. Measure of adaptiveness to environmental conditions, usually measured by success of offspring

food web. Set of organisms with interrelated food chains

forest regulation. Method to determine cutting patterns over the entire forest property

forestry. Art, science, and practice of studying and managing forests, plantations, and any other related natural forest resources

fragmentation. The level of discontinuities in a landscape. In a forestry landscape, fragmentation is measured by the amount of forest edges created by natural or human causes.

fuel management. Practices that manipulate vegetation to reduce fire hazards

Gaia theory. Theory that the whole earth is behaving somewhat like one individual organism

gap. Canopy opening in otherwise dense forest

group selection. Regeneration method that regenerates trees in groups, typical in uneven-aged forests

growing stock. Wood volume of living trees

hardwoods. Broadleaved trees

harvesting unit. Area in which trees are scheduled to be harvested in a single operation

herbivory. Predation where organisms eat plants

high-grading. A silvicultural practice that aims at removing only the most valuable trees without any consideration for the future quality of the forest

intermediate disturbance hypothesis. Theory that predicts highest diversity at intermediate disturbance levels

intermediate entry. Harvesting activities, such as thinning, not aimed at regeneration

island biogeography. The study that attempts to establish and explain the factors that affect species richness in any area surrounded by unsuitable areas such as deserts, lakes, clearcut, and mountains

landscape. Any visible features of an area of land that includes both its physical and biological elements

landscape ecology. A subdiscipline of ecology and geography that studies the effects of spatial variation in any particular landscape on ecological processes such as distribution of species, energy, and materials

late-successional forest. Forest that have undergone succession and developed without major disturbances for a long time

livestock. Animal kept by humans for commercial purposes, such as food or fiber

metapopulation. Groups of physically separated populations of the same species that interact among themselves

monoculture. Stand with a single species of trees

multivariate analysis. Statistical analysis of multiple variables at the same time

mutualism. Interaction between two species by which both species benefit

mycorrhizae. Symbiotic relationship between a fungus and plant root

natural disturbance. Natural forces that result in mortality of vegetation

neutral theory. Theory that assumes an individual plant's traits don't influence a plant's success

niche. Conditions along environmental gradients in which a species or population is found. In contrast to the fundamental niche, which encompasses the full range of conditions, realized niches are smaller because of plant interactions.

normal forest. Forest composed of even-aged, fully stocked stands representing a balance of age classes. Concept aimed at determining optimal harvest levels.

null hypothesis. Hypothesis that no statistical differences exist between samples

nursery. Operation to produce seedlings for outplanting, grown either in nursery fields (bareroot) or in greenhouses in containers (container nursery)

paradigm. Approach and underlying assumptions by which a discipline operates

partial cutting. Harvesting regime in which living trees are left behind

pest control. Management practices to minimize impact of damaging agents

plant association. Group of plant species usually found growing together

plantation. Forest established by planted seedlings

plant community. Collection of plants in the area that interact with each other

plant plasticity. The ability of a plant with a given genotype (genetic makeup) to change its phenotype (external features) in response to changes in the environment

population ecology. A subdivision of ecology that studies how various organisms of the same species interact with themselves and with their surrounding environment

productivity. A measure of output from a production process per unit of input. In forestry, productivity refers to the amount of wood or biomass produced per unit of time on a per-area basis.

q-factor. Ratio of trees in a size class to the number of trees in the next larger size class. Used to describe the reverse J-shaped diameter distribution curve in uneven-aged stands.

range of variability. Range of natural conditions in ecosystem composition, structure, and function

reductionism. Scientific view that assumes everything can be explained by interactions of smaller pieces

reforestation. Establishment of forest after harvesting or other disturbances

replication. Repetition of treatments in experiments to statistically determine variability associated with the treatment

resilience. Ability of an ecosystem to recover after disturbances

retention harvest. Harvest that retains living trees to benefit the next rotation

rotation. Time between regeneration harvests

scale (temporal or spatial). A relative measure of time or space.

scope of inference. Conditions reflected in study conditions and to which study results apply

seedbed. Substrate on which seeds are germinating

seed dispersal. Seed movement from tree to place of germination

seed rain. Quantity of seeds that fall per unit of area

seed tree cut. A silvicultural system that leaves some trees standing after clearcutting to allow the natural seeding of the cutover areas. There trees are often harvested after a few years.

selection cutting. A silvicultural system that removes only a small proportion of trees, usually the oldest or largest, either as single scattered trees or in small groups at relatively short intervals, commonly five to twenty years, typical for uneven-aged forests

shade tolerance. Ability of a plant to grow and survive in shade

shelterwood cutting. A silvicultural system that removes mature trees over a series of cuttings, which extend over a period of years. This is normally done to help the establishment of natural regeneration under the partial shelter of the trees left behind.

silvics. The study of how trees grow, reproduce, and respond to their environment

silvicultural prescription. Refers to a specific set of human interventions that are prescribed by a forester for a forest stand in order to achieve a certain silvicultural objective

silvicultural system. Refers to different approaches to harvesting, regenerating, and growing forests

silviculture. The art and science of producing and tending a forest to achieve the objectives of management

simulation model. A computer program that attempts to simulate a particular system in a dynamic way. In forestry, simulation models are used to predict the growth and yield of forest stands and their dynamics following natural or human-induced disturbance.

single-tree selection. *See* selection cutting

site. An area of land, especially with reference to its capacity to produce vegetation as a function of environmental factors, such as climate and soil

site index. A measure of the tree-growing quality of a forest site. It is based on the height (in feet or meters) that dominant trees will reach at a given age. This value is commonly expressed as a fifty-year site index. This measure is based on the observation that trees grow taller on richer sites independently of the stand density.

site preparation. Any treatment of a forest site to prepare the soil for the establishment of a new crop of trees by either plantation or natural means

slash. Residues of wood, branches, and leaves left following harvesting

spacing. Silvicultural intervention that removes a certain proportion of trees in a young or maturing stand to improve the growth of the remaining trees

speciation. Evolutionary process by which new species arise

sprouting. The ability of a tree to grow stems directly from its base, stump, or root. This is relatively common among hardwoods.

stand. Any aggregation of trees occupying a specific area in uniform enough composition (species), age, and arrangement to be distinguishable from the forest on adjoining areas

stand dynamics. Changes in species composition, structure, and function occurring in a forest stand over time

stand structure. Horizontal and vertical distribution of vegetation

stock type. Type of seedlings grown in a nursery, usually either bareroot in nursery soil or in containers

stocking. Number of trees in any particular stand. Usually expressed as a relative measure (well stocked/fully stocked, overstocked, understocked).

succession. The gradual replacement of one group of organisms by another over time following an initial disturbance

sustainability. Characteristic by which a process or state can be maintained at a certain level indefinitely

sustainable harvest level. Level of wood harvesting that can be sustained indefinitely. In forestry, this is often calculated as annual allowable cut on a per-year basis for any specific region.

sustained yield. Amount of a natural resource, such as wood, that can be extracted without reducing the inventory or production potential

taxonomy. Science of classifying plants and animals

thinning. Partial removal of trees in an immature stand to select for a specific species or to increase the growth rate of the remaining trees

tree taper. The gradual reduction of diameter in a stem of a tree or a log from the base to the top

TRIAD. A zoning allocation approach for any territory into three distinct zones. In forestry, TRIAD refers to the allocation into protected areas, intensive forest production areas, and extensive forest production areas.

underplanting. Plantation established beneath an already established overstory canopy

understory. Vegetation beneath an overstory canopy

uneven-aged stands. Stands composed of trees of multiple (ideally all) age classes

vegetation control. Removal of vegetation to improve growth of desired commercial tree species

References

Abetz, P., and J. Klädtke. 2000. Die Df-2000—Eine Entscheidungshilfe für Durchforstungen. *AFZ-Der Wald* 55:454–55.

Alemdag, I. S. 1978. Evaluation of some competition indexes for the prediction of diameter increment in planted white spruce. *Canadian Forest Service Information Report* FMR-X-108.

Amateis, R. L., P. J. Radtke, and H. E. Burkhart. 1995. TAUYIELD: A stand-level growth and yield model for thinned and unthinned loblolly pine plantations. *Loblolly Pine Growth and Yield Cooperative* Report No. 82.

Ammon, W. 1955. *Das Plenterprinzip in der Waldwirtschaft*. Bern, Switzerland: Verlag Paul Haupt.

Anderson, P.W. 1972. More is different. *Science* 177:393–96.

Andren, H. 1994. Effects of habitat fragmentation on birds and mammals in landscapes with different proportions of suitable habitat: A review. *Oikos* 71:355–66.

Angelstam, P. K. 1998. Maintaining and restoring biodiversity in European boreal forests by developing natural disturbance regimes. *Journal of Vegetation Science* 9:593–602.

Angers, V. A., C. Messier, M. Beaudet, and A. Leduc. 2005. Comparing composition and structure between old-growth and selectively harvested stands in hardwood forest eastern Canada. *Forest Ecology Management* 217:275–93.

Anonymous. 2005. *Jahresbilanz 2004: Rückblick für die Zukunft*. Baden-Württemberg, Ministerium für Ernährung und ländlichen Raum. Stuttgart, Germany. http://www.wald-online-bw.de/fileadmin/lfv_pdf/jahresbilanz/Jahresbilanz_2004.pdf. Accessed Oct. 20, 2007.

Arnott, J. T., and W. J. Beese. 1997. Alternatives to clearcutting in B.C. coastal montane forests. *Forest Chronicles* 73:670–78.

Arthur, B. 1999. Complexity and economy. *Science* 284:107–9.

Assmann, E. 1961. *Waldertragskunde: Organische Produktion, Struktur, Zuwachs und Ertrag von Waldbestanden*. Munich, Germany: BLV Verlagsgesellschaft.

Attiwill, P. M. 1994. The disturbance of forest ecosystems: The ecological basis for conservative management. *Forest Ecology and Management* 63:247–300.

Attiwill, P. M., and M. A. Adams. 1993. Tansley Review No. 50. Nutrient cycling in forests. *New Phytologist* 124:561–82.

Aubin, I., S. Gachet, C. Messier, and A. Bouchard. 2007. How resilient are northern hardwood forests to human disturbance? An evaluation using a plant functional group approach. *Ecoscience* 14:259–71.

Aubry, K. B., M. P. Amaranthus, C. B. Halpern, J. D. White, B. L. Woodard, C. E. Peterson, C. A. Lagoudakis, and A. J. Horton. 1999. Evaluating the effects of varying levels and patterns of green-tree retention: Experimental design of the DEMO study. *Northwest Science* 73:12–26.

Backéus, S., L. O. Eriksson, and F. Garcia. 2005. Impact of climate change uncertainty on optimal forest management policies at stand level. In *MODSIM 2005 International Congress on Modelling and Simulation*, ed. A. Zerger and R. M. Argent, 468–74. Australia: Modelling and Simulation Society of Australia and New Zealand.

Barbour, M. 1996. American ecology and American culture in the 1950s: Who led whom? *Bulletin of the Ecological Society of America* 77:44–51.

Barnes, B. V., D. R. Zak, S. R. Denton, and S. H. Spurr. 1998. *Forest ecology, 4th edition*. New York: Wiley.

Bartemucci, P., C. Messier, and C. D. Canham. 2006. Overstory influences on light attenuation patterns and understory plant diversity in southern boreal forests of Quebec. *Canadian Journal of Forest Research* 36:2065–79.

Bazzaz, F., G. Ceballos, M. Davis, R. Dirzo, P. Ehrlich, R. T. Eisner, S. Levin, J. H. Lawton, J. Lubchenco, P. A. Matson, H. A. Mooney, P. H. Raven, J. E. Roughgarden, J. Sarukhan, D. Tilman, P. Vitousek, B. Walker, D. H. Wall, E. O. Wilson, and G. M. Woodwell. 1998. Ecological science and the human predicament. *Science* 282:879.

Beaudet, M., C. Messier, and A. Leduc. 2004. Temporal variation in light availability and understory recovery following selection cutting in northern hardwood stands. *Journal of Ecology* 92:328–38.

Bédard, S., and Brassard, F. 2002. Les effets réels des coupes de jardinage dans les forêts publiques du Québec en 1995 et 1996. Ministère des Ressources naturelles, Gouvernement du Québec 2002–3117.

Begon, M., C. Townsend, and J. L. Harper. 2006. *Ecology: From individuals to ecosystems, 4th edition*. Malden, MA: Blackwell.

Bell, G. 2000. The distribution of abundance in neutral communities. *American Naturalist* 155:606–17.

Benecke, U. 1996. Ecological silviculture: The application of age-old methods. *New Zealand Forestry* 41:27–33.

Bengtsson, J., P. Angelstam, T. Elmgvist, U. Emanuelsson, C. Folke, M. Ihse, F. Mobert, and M. Nystrom. 2003. Reserves, resilience and dynamic landscapes. *Ambio* 32:389–96.

Bengtsson, J., S. G. Nilsson, A. Franc, and P. Menozzi. 2000. Biodiversity, disturbances, ecosystem function and management of European forests. *Forest Ecology and Management* 132:39–50.

Benzie, J.W. 1977. *Manager's handbook for red pine in the north-central states.* St. Paul, MN: U.S. Dept. of Agriculture, Forest Service, North Central Forest Experiment Station, General Technical Report NC-33.

Berger, A. L., and K. J. Puettmann. 2000. Overstory composition and stand structure influence herbaceous plant diversity in the mixed aspen forest of northern Minnesota. *American Midland Naturalist* 143:111–25.

Bergeron, Y., and B. Harvey. 1997. Basing silviculture on natural ecosystem dynamics: an approach applied to the southern boreal mixedwood forest of Quebec. *Forest Ecology and Management* 92:235–42.

Bergeron, Y., B. Harvey, A. Luduc, and S. Gauthier. 1999a. Forest management guidelines based on natural disturbance dynamics: Stand- and forest-level considerations. *The Forestry Chronicle* 75:49–54.

Bergeron, Y., B. Harvey, A. Luduc, and S. Gauthier. 1999b. Forest management strategies based on the dynamics of natural disturbances: Considerations and a proposal for a model allowing an even-management approach. *Forestry Chronicle* 75:55–61.

Bergeron, Y., A. Leduc, B. D. Harvey, and S. Gauthier. 2002. Natural fire regime: A guide for sustainable management of the Canadian boreal forest. *Silva Fennica* 36:81–95.

Biolley, H. 1920. L'aménagement des forêts par la méthode expérimentale et spécialement la méthode du contrôle. Translated into German by Eberbach 1922.

Borsuk, M. E., C. A. Stow, and K. H. Reckhow. 2004. A Bayesian network of eutrophication models for synthesis, prediction, and uncertainty analysis. *Ecological Modelling* 173:219–39.

Botkin, D., ed. 2002. *Forces of change: A new view of nature.* New York: Simon & Schuster.

Botkin, D. B., J. F. Janak, and J. R. Wallis. 1972a. Rationale, limitations, and assumptions of a Northeastern Forest Growth Simulator. *IBM Journal of Research and Development* 16:101–16.

Botkin, D. B., J. F. Janak, and J. R. Wallis. 1972b. Some ecological consequences of a computer model of forest growth. *Journal of Ecology* 60:849–72.

Bottom, D. L., G. H. Reeves, and M. H. Brookes, eds. 1996. *Sustainability issues for resource managers.* Portland, OR: U.S. Department of Agriculture, Forest Service, Pacific Northwest Research Station, General Technical Report PNW-GTR-370.

Boucher, D., L. Degranpré, and S. Gauthier. 2003. Développement d'un outil de classification de la structure des peuplements et comparaison de deux territoires de la pessière à mousses du Québec. *Forestry Chronicle* 79:318–28.

Bradbury, R. H., J. D. Van Der Laan, and D. G. Green. 1996. The idea of complexity in ecology. *Senckenbergiana marit* 27:89–96.

Brais, S., B. D. Harvey, Y. Bergeron, C. Messier, D. Greene, A. Belleau, and D. Paré. 2004. Testing forest ecosystem management in boreal mixedwoods of northwestern Quebec: Initial response of aspen stands to different levels of harvesting. *Canadian Journal of Forest Research* 34:431–46.

Brang, P. 2007. Wer glaubt, weiss mehr: Die Forstbranche zwischen Tradition und Innovation. http://www.forest.ch/meinung/downloads/opi_50_brang.pdf. Accessed Nov. 12, 2007.

Braun-Blanquet, J. 1928. *Pflanzensoziologie: Grundzüge der Vegetationskunde.* Vienna, Austria: Springer.

Bray, J. R., and J. T. Curtis. 1957. An ordination of the upland forest communities of southern Wisconsin. *Ecological Monographs* 27:325–49.

Brazee, R. J. 2001. Introduction—The Faustmann Formula: Fundamental to forest economics 150 years after publication. *Forest Science* 47:441–42.

Breckling, B., F. Müller, H. Reuter, F. Hölker, and O. Fränzle. 2005. Emergent properties in individual-based ecological models: Introducing case studies in an ecosystem research context. *Ecological Modelling* 186:376–88.

Briggs, D., and J. Trobaugh. 2001. *Management practices on Pacific Northwest west-side industrial forest lands, 1991–2000: With projections to 2005.* Seattle: College of Forest Resources, University of Washington, Stand Management Cooperative Working Paper No 2.

Brokaw, N. V. L. 1985. Gap-phase regeneration in a tropical forest. *Ecology* 66:682–87.

Brooker, R. W. 2006. Plant-plant interactions and environmental change. *New Phytologist* 171:271–84.

Bruce, D. 1977. Yield differences between research plots and managed forests. *Journal of Forestry* 75:14–17.

Brumelle, S., D. Granot, M. Halme, and I. Vertinsky. 1998. A tabu search algorithm for finding good forest harvest schedules satisfying green-up constraints. *European Journal of Operational Research* 106:408–24.

Buckman, R. E. 1962. *Growth and yield of red pine in Minnesota.* U.S. Department of Agriculture, Forest Service, Technical Bulletin 1272.

Buergi, M., and A. Schuler. 2003. Driving forces of forest management: An analysis of regeneration practices in the forests of the Swiss Central Plateau during the 19th and 20th century. *Forest Ecology and Management* 176:173–83.

Bunnell, F. L., L. L. Kremsater, and E. Wind. 1999. Managing to sustain vertebrate richness in forests of the Pacific Northwest: Relationships within stands. *Environmental Reviews* 7:97–146.

Burkhart, H. E., K. D. Farrar, R. L. Amateis, and R. F. Daniels. 2001. *Simulation of individual tree growth and stand development in loblolly pine plantations on cutover, site-prepared areas.* Blacksburg, VA: Department of Forestry, Virginia. Technical Publication No. FWS-1-87.

Burnham, K. P., and D. R. Anderson. 2002. *Model selection and multimodel inference: A practical information-theoretic approach, 2nd edition*. New York: Springer.

Burns, R. M., and B. H. Honkala, eds. 1990. *Silvics of North America: 1. conifers; 2. hardwoods*. Washington, DC: U.S. Department of Agriculture, Forest Service Agriculture Handbook 654.

Burschel, P., and J. Huss. 1997. *Grundriss des Waldbaus*. Berlin, Germany: Blackwell Wissenschafts-Verlag.

Burton, P. J, C. Messier, D. W. Smith, and W. L. Adamovicz, eds. 2003. *Toward sustainable management of boreal forest: Emulating nature, minimizing impacts and supporting communities*. Ottawa, Canada: NRC Press.

Canham, C. D., P. LePage, and K. D. Coates. 2004. A neighbourhood analysis of canopy tree competition: Effects of shading versus crowding. *Canadian Journal of Forest Research* 34:778–87.

Canham, C. D., and M. Uriarte. 2006. Analysis of neighborhood dynamics of forest ecosystems using likelihood methods and modeling. *Ecological Applications* 16:62–73.

Cannell, M. G. R., D. C. Malcolm, and P. A. Robertson, eds. 1992. *The ecology of mixed-species stands of trees*. Oxford, UK: Blackwell Scientific.

Carroll, A. L., S.W. Taylor, J. Régnière, L. Safranyik. 2004. Effects of climate change on range expansion by the mountain pine beetle in British Columbia. In *Mountain pine beetle symposium: Challenges and solutions*, ed. T. L. Shore, J. E. Brooks, and J. E. Stone, 223–32. Information Report BC-X-399. Natural Resources Canada, Canadian Forest Service, Pacific Forestry Centre, Victoria, British Columbia.

Catovsky, S., and F. A. Bazzaz. 2000. The role of resource interactions and seedling regeneration in maintaining a positive feedback in hemlock stands. *Journal of Ecology* 88:100–12.

Chapin, F. S. III, E. S. Zavaleta, V. T. Eviner, R. L. Naylor, P. M. Vitousek, H. L. Reynolds, D. U. Hooper, W. K. Lauenroth, A. Lombard, H. A. Mooney, A. R. Mosier, S. Naeem, S. W. Pacala, J. Roy, W. L. Steffen, and D. Tilman. 2000. Consequences of changing biodiversity. *Nature* 405:234–42.

Chase, J. M., and M. A. Leibold. 2003. *Ecological niches: Linking classical and contemporary approaches*. Chicago: University of Chicago Press.

Chave, J. 2004. Neutral theory and community ecology. *Ecology Letters* 7:241–53.

Childs, S. W., and L. E. Flint. 1987. Effect of shadecards, shelterwoods, and clearcuts on temperature and moisture environments. *Forest Ecology and Management* 18:205–17.

Cissel, J. H., P. D. Anderson, S. Berryman, S. S. Chan, D. H. Olson, K. J. Puettmann, and C. Thompson. 2006. *BLM density management and riparian buffer study: Establishment report and study plan*. Reston, VA: U.S. Geological Survey, Scientific Investigations Report 2006-5087.

Clark, J. S., E. Macklin, and L. Wood. 1998. Stages and spatial scales of recruitment limitation in southern Appalachian forests. *Ecological Monographs* 68:213–35.

Cleary, B. D., R. D. Greaves, and R. K. Hermann. 1978. *Regenerating Oregon's forests:*

A guide for the regeneration forester. Corvallis, OR: Oregon State University Extension Service.

Clements, F. E. 1905. *Research methods in ecology*. Lincoln, NE: Jacob North & Company.

Clements, F. E. 1936. Nature and structure of the climax. *Journal of Ecology* 24:252–84.

Coates, K. D. 1997. Windthrow damage two years after partial cutting of the Date Creek silvicultural systems study in the interior cedar-hemlock forests of northwestern British Columbia. *Canadian Journal of Forest Research* 27:1695–1701.

Coates, K. D. 2000. Conifer seedling response to northern temperate forest gaps. *Forest Ecology and Management* 127:249–69.

Coates, K. D. 2002. Tree recruitment in gaps of various size, clearcuts and undisturbed mixed forest of interior British Columbia, Canada. *Forest Ecology and Management* 155:387–98.

Coates, K. D., A. Banner, J. D. Steventon, P. LePage, and P. Bartemucci. 1997. *The Date Creek silvicultural systems study in the interior cedar-hemlock forests of northwestern British Columbia: Overview and treatment summaries.* Land Management Handbook 38. Victoria, Canada: British Columbia Ministry of Forests.

Coates, K. D., and P. J. Burton, 1997. A gap-based approach for development of silvicultural systems to address ecosystem management objectives. *Forest Ecology and Management* 99:337–54.

Coates, K. D, and P. J. Burton. 1999. Growth of planted tree seedlings in response to ambient light levels in northwestern interior cedar-hemlock forests of British Columbia. *Canadian Journal of Forest Research* 29:1374–82.

Coates, K. D., C. D. Canham, M. Beaudet, D. L. Sachs, and C. Messier. 2003. Use of a spatially explicit individual-tree model (SORTIE/BC) to explore the implications of patchiness in structurally complex forests. *Forest Ecology and Management* 186:297–310.

Coates, K. D., and J. D. Steventon. 1995. Patch retention harvesting as a technique for maintaining stand level biodiversity in forests of north central British Columbia. In *Innovative silvicultural systems in boreal forests*, ed. C. R. Bamsey, 102–6. Edmonton, Canada: Clear Lake Ltd.

Connell, J. H. 1978. Diversity in tropical rain forests and coral reefs. *Science* 199:1302–10.

Connell, J. H. 1980. Diversity and coevolution of competitors, or the ghost of competition past. *Oikos* 35:131–38.

Cornett, M. W., K. J. Puettmann, and P. B. Reich. 1998. Canopy type, leaf litter, predation, and competition influence conifer regeneration and early survival in two Minnesota conifer-deciduous forests. *Canadian Journal of Forest Research* 28:196–205.

Corona, P., and A. Ferrara. 1989. Individual competition indices for conifer plantations. *Agriculture, Ecosystems and Environment* 27:429–37.

Cotta, H. 1816. Cotta's preface. Forest history today (2000). Reprinted from *Forestry Quarterly* 1, 1902–1903: 27–28.

Cotta, H. 1817. *Anweisung zum Waldbau.* Dresden, Germany: Arnold Verlag.

Courbaud, B., F. Goreaud, P. H. Dreyfus, and F. R. Bonnet. 2001. Evaluating thinning strategies using a tree distance dependent growth model: some examples based on the CAPSIS software uneven-aged spruce forests module. *Forest Ecology and Management* 145:15–28.

Couteron, P., R. Pelissier, E. A. Nicolinie, and D. Paget. 2005. Predicting tropical forest stand structure parameters from Fourier transform of very high-resolution remotely sensed canopy images. *Journal of Applied Ecology* 42:1121–28.

Crow, T. R., D. S. Buckley, E. A. Nauertz, and J. C. Zasada. 2002. Effects of management on the composition and structure of northern forests in upper Michigan. *Forest Science* 48:129–45.

Csada, R. D., P. C. James, and R. H. M. Espie. 1996. The "file drawer problem" of non-significant results: Does it apply to biological research? *Oikos* 76:591–93.

Curtis, R. O. 1998. "Selective cutting" in Douglas-fir: History revisited. *Journal of Forestry* 96:40–46.

Curtis, R. O., D. S. DeBell, R. E. Miller, M. Newton, J. B. St. Clair, and W. I. Stein. 2007. *Silvicultural research and the evolution of forest practices in the Douglas-fir region.* U.S. Department of Agriculture, Forest Service General Technical Report PNW-GTR-696.

Curtis, R. O., D. D. Marshall, and D. S. DeBell, eds. 2004. *Silvicultural options for young-growth Douglas-fir forests: The Capitol Forest Study—Establishment and first results.* U.S. Department of Agriculture, Forest Service General Technical Report PNW-GTR-598.

Daily, G. C. 1997. *Nature's services: Societal dependence on natural ecosystems.* Washington, DC: Island Press.

D'Amato, A. W., and K. J. Puettmann. 2004. The relative dominance hypothesis explains interaction dynamics in mixed species Alnus rubra/Pseudotsuga menziesii stands. *Journal of Ecology* 92:450–63.

Daniel, T. W., J. A. Helms, and F. S. Baker. 1979. *Principles of silviculture, 2nd edition.* New York: McGraw-Hill.

Daniels, R. F. 1976. Simple competition indices and their correlation with annual loblolly pine tree growth. *Forest Science* 22:454–56.

Daniels, R. F., and H. E. Burkhart. 1975. *Simulation of individual tree growth and stand development in managed loblolly pine plantations.* Blacksburg, VA: Division of Forestry and Wildlife Research, Virginia. Polytechnical Institute and State University Publication: FWS-5-75.

Daniels, R. F., H. E. Burkhart, and T. R. Clason. 1986. A comparison of competition measures for predicting growth of loblolly pine trees. *Canadian Journal of Forest Research* 16:1230–37.

D'Antonio, C. M., J. T. Tunison, and R. K. Loh. 2000. Variation in the impact of exotic grasses on native plant composition in relation to fire across an elevation gradient in Hawaii. *Austral Ecology* 25:507–22.

Davis, L. S., K. N. Johnson, P. S. Bettinger, and T. E. Howard. 2001. *Forest management: To sustain ecological, economic, and social values, 4th edition.* New York: McGraw-Hill.

Dean, T. J. 2004. Basal area increment and growth efficiency as functions of canopy dynamics and stem mechanics. *Forest Science* 50:106–16.

Delic, K. A., and R. Dum. 2006. *On the emerging future of complexity sciences*. http://www.acm.org/ubiquity/views/v7i10_complexity.html. Accessed Apr. 27, 2008.

de Liocourt, F. 1898. De l'aménagement des sapinières. *Bulletin trimestriel, Société forestière de Franche-Comtéet Belfort* 4:396–409.

de Montigny, L. 2004. *Silviculture treatments for ecosystem management in the sayward (STEMS): Establishment report for STEMS 1*. Victoria, Canada: Snowden Demonstration Forest, Research Branch, British Columbia Ministry of Forests, Technical Report 17.

Dengler, A. 1930. *Waldbau auf ökologischer Grundlage*. Berlin, Germany: Julius Springer.

Denslow, J. S., and G. S. Hartshorn. 1994. Tree-fall gap environments and forest dynamic processes. In *La selva: Ecology and natural history of a neotropical rain forest*, ed. L. A. McDade, K. S. Bawa, H. A. Hespenheide, and G. S. Hartshorn, 120–27. Chicago: University of Chicago Press.

Diamond, J. 1999. *Guns, germs, and steel: The fates of human societies*. New York: Norton.

Doig, I. 1976. The murky annals of clearcutting: A 40-year-old dispute. *Pacific Search* 10:12–14.

Drever, C. R., G. Peterson, C. Messier, Y. Bergeron, and M. Flannigan. 2006. Can forest management based on natural disturbances maintain ecological resilience? *Canadian Journal of Forest Research* 36:2285–99.

du Monceau, D. 1766. *Von der Fällung der Wälder und gehöriger Anwendung des gefällten Holzes oder wei mit Schlagholz . . . umzugehen*. (Translated from French by C. Chr. von ölhafen von Schöllenbach.) Nürnberg, Germany.

Edwards, K. S., and K.J. Kirby. 1998. The potential for developing a normal age-structure in managed ancient woodland at a local scale in three English counties. *Forestry* 71:365–71.

Ehrlich, P. R., and P. H. Raven. 1964. Butterflies and plants: A study in coevolution. *Evolution* 18:586–608.

Elton, C. S. 1927. *Animal ecology*. London: Sidgwick & Jackson.

Emmeche, C. 1997. Aspects of complexity in life and science. *Philosophica* 59:41–68.

Ericksson, L. O. 2006. Planning under uncertainty at the forest level: A systems approach. *Scandinavian Journal of Forest Research* 21:111–17.

Fahey, R., and K. J. Puettmann. 2007. Ground-layer disturbance and initial conditions influence gap partitioning of understory vegetation. *Journal of Ecology* 95:1098–1109.

Fargione, J., C. S. Brown, and D. Tilman. 2003. Community assembly and invasion: An experimental test of neutral versus niche processes. *Proceedings of the National Academy of Sciences of the United States of America* 100:8916–20.

Farrell, E. P., R. D. Führer, F. Anderson, R. Hüttl, and P. Piussi. 2000. European forest ecosystems: Building the future on the legacy of the past. *Forest Ecology and Management* 132:5–20.

Faustmann, M. 1849. Berechnung des Werthes, welchen Waldboden, so wie noch nicht haubare Holzbestande für die Waldwirtschaft besitzen (On the determination of the value which forestland and immature stands pose for forestry). *Allgemeine Forst und Jagdzeitung* 15. (Reprinted in *Journal of Forest Economics,* 1995, 1:7–44.)

Folke, C., S. Carpenter, B. Walker, M. Scheffer, T. Elmqvist, L. Gunderson, and C. S. Holling. 2004. Regime shifts, resilience, and biodiversity in ecosystem management. *Annual Review Ecology and Evolution* 35:557–81.

Food and Agriculture Organization. 1997. *Issues and opportunities in the evolution of private forestry and forestry extension in several countries with economies in transition in central and eastern Europe.* http://www.fao.org/docrep/w7170E/w7170e00.htm. Accessed June 6, 2008.

Foster, E. 1952. Approved logging technique. *Journal of Forestry* 50:135–36.

Franklin, J. F., D. Berg, D. A. Thornburgh, and J. C. Tappeiner. 1997. Alternative silvicultural approaches to timber harvesting: Variable retention harvest systems. In *Creating a forestry for the 21st century: The science of ecosystem management,* ed. K. A. Kohm and J. F. Franklin, 111–39. Washington, DC: Island Press.

Franklin, J. F., and D. Lindenmayer, eds. 2003. *Towards forest sustainability.* Washington, DC: Island Press.

Franklin, J. F., D. Lindenmayer, J. A. MacMahon, A. McKee, J. Magnuson, D. A. Perry, R. Waide, and D. Foster. 2000. Threads of continuity. *Conservation Biology in Practice* 1:9–16.

Franklin, J. F., T. A. Spies, R. Van Pelt, A. B. Carey, D. A. Thornburgh, D. R. Berg, D. B. Lindenmayer, M. E. Harmon, W. S. Keeton, D. C. Shaw, K. Bible, and J. Chen. 2002. Disturbances and structural development of natural forest ecosystems with silvicultural implications, using Douglas-fir forests as an example. *Forest Ecology and Management* 155:399–423.

Freise, C. 2007. Fakten statt forstlicher Götterblick: Wie das Einzelbaumwachstum der Fichte über die relative Kronenlänge gesteuert werden kann. *Forst und Holz* 62:31–34.

Frelich, L. E. 2002. *Forest dynamics and disturbance regimes: Studies from temperate evergreen-deciduous forests.* Cambridge, UK: Cambridge University Press.

Frelich, L. E., and K. Puettmann. 1999. Restoration ecology. In *Maintaining biodiversity in forest ecosystems,* ed. M. L. Hunter, Jr., 498–524. Cambridge, UK: Cambridge University Press.

Fries, C., O. Johansson, B. Pettersson, and P. Simonsson. 1997. Silvicultural models to maintain and restore natural stand structures in Swedish boreal forests. *Forest Ecology and Management* 94:89–103.

Fujimori, T. 2001. *Ecological and silvicultural strategies for sustainable forest management.* Amsterdam, The Netherlands: Elsevier Science.

Gallagher, R., and T. Appenzeller. 1999. Beyond reductionism. *Science* 284:79.

Gamborg, C., and J. B. Larsen. 2003. Back to nature: A sustainable future for forestry? *Forest Ecology and Management* 179:559–71.

Ganio, L., and K. J. Puettmann. 2008. Challenges in statistical inference for large operational experiments. *Journal of Sustainable Forestry* 26:1–18.

Gause, G. F. 1934. *The struggle for existence.* Baltimore: Williams & Wilkins.

Gauthier, S., L. DeGranpré, and Y. Bergeron. 2000. Differences in forest composition in two boreal forest ecoregions of Québec. *Journal of Vegetation Science* 11:781–90.

Gayer, K. 1880. *Der Waldbau.* Berlin, Germany: Wiegandt & Hempel & Parey.

Gayer, K. 1886. *Der gemischte Wald, seine Begründung und Pflege, insbesondere durch Horst-und Gruppenwirtschaft.* Berlin, Germany: Parey Verlag.

Gershenson, C., and F. Heylighen. 2003. When can we call a system self-organizing? In *Advances in artificial life*, ed. W. Banzhaf, T. Christaller, P. Dittrich, J. T. Kim, and J. Ziegler, 606–14. 7th European Conference, ECAL Dortmund. Germany: Springer.

Gilbert, B., and M. J. Lechowicz. 2004. Neutrality, niches, and dispersal in a temperate forest understory. *Proceedings of the National Academy of Sciences of the United States of America* 101:7651–56.

Gleason, H. A. 1926. The individualistic concept of the plant association. *Bulletin of the Torrey Botanical Club* 53:7–26.

González, J. M., M. P. Nicolau, and P.V. Grau. 2006. *Manual de ordenación por rodales: gestión multifunctional de los espacios forestales.* Solsona, Spain: Centre Tecnològic Forestal de Catalunya.

Goodall, D. W. 1954. Objective methods for the classification of vegetation III: An essay in the use of factor analysis. *Australian Journal of Botany* 2:304–24.

Graham, R. T., and T. B. Jain. 2004. Past, present, and future role of silviculture in forest management. In *Silviculture in special places: Proceedings of the 2003 National Silviculture Workshop*, ed. W. D. Shepperd and L. G. Eskew, 1–14. Fort Collins, CO: U.S. Department of Agriculture, Forest Service. Proceedings RMRS-P-34.

Gratzer, G., C. Canham, U. Dieckmann, A. Fischer, Y. Iwasa, R. Law, M. J. Lexer, H. Sandmann, T. A. Spies, B. E. Splechtna, and J. Szwagrzyk. 2004. Spatio-temporal development of forests: Current trends in field methods and models. *Oikos* 107:3–15.

Gravel, D., C. Canham, M. Beaudet, and C. Messier. 2006. Reconciling niche and neutrality: The continuum hypothesis. *Ecology Letters* 9:399–409.

Green, J. L., A. Hasting, P. Arzberger, F. J. Ayala, K. L. Cottingham, K. Cuddington, F. Davis, J. A. Dunne, M.-J. Fortin, L. Gerber, and M. Neubert. 2005. Complexity in ecology and conservation: Mathematical, statistical and computational challenges. *Bioscience* 55:501–10.

Greene, D. F., C. D. Canham, K. D. Coates, and P. T. LePage. 2004. An evaluation of alternative dispersal functions for trees. *Journal of Ecology* 92:758–66.

Guldin, J. M. 2004. Overview of ecosystem management research in the Ouachita and Ozark Mountains: Phases I–III. In *Ouachita and Ozark mountains symposium: Ecosystem management research*, ed. J. M. Gulding, 8–14. U.S. Department of Agriculture, Forest Service, Southern Research Station, General Technical Report SRS-74.

Gunderson, L. H. 2000. Ecological resilience: In theory and application. *Annual Review of Ecology and Systematics* 31:425–39.

Gunderson, L. H., and C. S. Holling. 2002. *Panarchy: Understanding transformations in human and natural systems.* Washington, DC: Island Press.

Gunton, R. M., and W. E. Kunin. 2007. Density effects at multiple scales in an experimental plant population. *Journal of Ecology* 95:435–45.

Hampe, M. 2003. *Von der hierarchischen Welt zur homogenen Natur: Einführung in die Geschichte der Kosmologie.* Skript zur Vorlesung Erstellt von Sascha Jürgens, M.A. (Stand Wintersemester 03/04). ETH Zürich. http://www.phil.ethz.ch/education/SkriptKosmologie.pdf. Accessed Sept. 3, 2007.

Hann, W. J., M. A. Hemstrom, R. W. Haynes, J. Clifford, and R. A. Gravenmier. 2001. Costs and effectiveness of multi-scale integrated management. *Forest Ecology and Management* 153:127–45.

Hansen, A. J., T. A. Spies, F. J. Swanson, and J. L. Ohmann. 1991. Conserving biodiversity in managed forests: Lessons from natural forests. *BioScience* 41:382–92.

Hanski, I. 1999. *Metapopulation ecology.* Oxford, UK: Oxford University Press.

Hanzlik, E. J. 1922. Determination of the annual cut on a sustained basis for virgin American forests. *Journal of Forestry* 20:611–25.

Harper, J. L. 1967. A Darwinian approach to plant ecology. *Journal of Applied Ecology* 4:267–90.

Harper, J. L. 1977. *Population biology of plants.* London: Academic Press.

Harper, J. L. 1982. After description. In *The plant community as a working mechanism,* ed. E. I. Newman, 11–25. Oxford: Blackwell Scientific.

Hartig, G. L. 1791. *Anweisung zur Holzzucht für Förster.* Marburg, Germany.

Hartig, G. L. 1795. *Anweisung zur Taxation der Forste oder zu Bestimmung des Holzertrages der Wälder.* Gießen, Germany.

Harvey, B. D., A. Leduc, S. Gauthier, and Y. Bergeron. 2002. Stand-landscape integration in natural disturbance–based management of the southern boreal forest. *Forest Ecology and Management* 155:369–85.

Hasel, K. 1985. *Forstgeschichte.* Hamburg, Germany: Parey Verlag.

Hatzfeld, H. G. 1995. *Ökologische Waldwirtschaft.* Heidelberg, Germany: Verlag C.F. Muller GmbH.

Hausrath, H. 1982. *Geschichte des deutschen Waldbaus: Von seinen Anfängen bis 1850.* Freiburg, Germany: Hochschulverlag.

Hawley, R. C. 1921. *The practice of silviculture.* New York: Wiley.

Hawley, R. C., and D. M. Smith. 1972. *Silvicultura práctica.* Trad. del Inglés por Journal. Terradas 6a. Ed. Barcelona. Omega.

Helms, J. A. 1998. *Dictionary of forestry.* Bethesda, MD: Society of American Foresters.

Heyer, C. 1841. *Die Waldertragsregelung.* Giessen, Germany.

Hilborn, R., and M. Mangel. 1997. *The ecological detective: Confronting models with data.* Princeton, NJ: Princeton University Press.

Hilborn, R., C. J. Walters, and D. Ludwig. 1995. Sustainable exploitation of renewable resources. *Annual Review of Ecology and Systematics* 26:45–67.

Hobbs, R. J., and R. Hilborn. 2006. Alternatives to statistical hypothesis testing in ecology: A guide to self teaching. *Ecological Applications* 16:5–19.

Hobbs, N. T., S. Twombly, and D. S. Schimel. 2006. Invited feature: Deepening ecological insights using contemporary statistics. *Ecological Applications* 16:3–117.

Høiland, K., and E. Bendiksen. 1996. Biodiversity of wood-inhabiting fungi in a boreal coniferous forest in Sør-Trøndelag County. *Central Norway Nordic Journal of Botany* 16:643–59.

Holling, C. S. 1973. Resilience and stability of ecological systems. *Annual Review of Ecology and Systematics* 4:1–23.

Holling, C. S. 1992. Cross-scale morphology, geometry and dynamics of ecosystems. *Ecological Monographs* 62:447–502.

Holling, C. S., and G. K. Meffe. 1996. Command and control and the pathology of natural resources management. *Conservation Biology* 10:328–37.

Holmes, M. J., and D. D. Reed. 1991. Competition indices for mixed-species northern hardwoods. *Forest Science* 37:1338–49.

Hooper, D. U., F. S. Chapin III, J. J. Ewel, A. Hector, P. Inchausti, S. Lavorel, J. H. Lawton, D. M. Lodge, M. Loreau, S. Naeem, B. Schmid, H. Setälä, A. J. Symstad, J. Vandermeer, and D. A. Wardle. 2005. Effects of biodiversity on ecosystem functioning: A consensus of current knowledge. *Ecological Monographs* 75:3–35.

Houle, G. 1998. Seed dispersal and seedling recruitment of *Betula alleghaniensis*: Spatial inconsistency in time. *Ecology* 79:807–18.

Hubbell, S. P. 1997. A unified theory of biogeography and relative species abundance and its application to tropical rain forests and coral reefs. *Coral Reefs* 16, Suppl.: S9–S21.

Hubbell, S. P. 2001. *The Unified Neutral Theory of biodiversity and biogeography*. Princeton, NJ: Princeton University Press.

Hummel, S., and R. J. Barbour. 2007. Landscape silviculture for late-successional reserve management. In *Restoring fire-adapted ecosystems: Proceedings of the 2005 national silviculture workshop*, ed. Robert F. Powers, 157–69. Albany, CA: Pacific Southwest Research Station, Forest Service, U.S. Department of Agriculture.

Hundeshagen, J. C. 1826. *Die Forstabschatzung auf neuen wissenschaftlichen Grundlagen*. Tubingen, Germany: H. Laupp.

Hunter, M. L. 1990. *Wildlife, forests, and forestry: Principles of managing forests for biological diversity*. Englewood Cliffs, NJ: Prentice-Hall.

Hunter, M. L. 1999. *Maintaining biodiversity in forest ecosystems*. Cambridge, UK: Cambridge University Press.

Huston, M. A., L. W. Aarssen, M. P. Austin, B. S. Cade, J. D Fridley, E. Garnier, J. P. Grime, J.Hodgson, W. K. Lauenroth, K. Thompson, J. H. Vandermeer, and D. A. Wardle. 2000. No consistent effect of plant diversity on productivity. *Science* 289:1255.

Hutchinson, G. E. 1957. Concluding remarks. Population studies: Animal ecology and demography. *Cold Spring Harbor Symposia on Quantitative Biology* 22:415–27.

Isaac, L. A., R. S. Walters, D. M. Smith, and R. A. Brandes. 1952. Forest practices based on facts, not fancy. *Journal of Forestry* 50:562–65.

Jakobsen, M. K. 2001. History and principles of close to nature forest management: A central European perspective. *Textbook 2—Tools for Preserving Woodland Biodiversity*,

Nature Conservation Experience Exchange, Naconex: 56–60. http://www.pronatura.net/naconex/news5/E2_11.pdf. Accessed Jan. 23, 2008.

Johnson, E. A., and K. Miyanishi, eds. 2007. *Plant disturbance ecology: The process and the response.* Oxford, UK: Elsevier Science.

Johnson, J. B., and K. S. Omland. 2004. Model selection in ecology and evolution. *Trends in Ecology and Evolution* 19:101–8.

Kangur, A., ed. 2004. Natural disturbances and ecosystem based forest management. *Proceedings of the International Conference.* Tartu, Estonia: Transactions of the Faculty of Forestry, Estonian Agricultural University (No. 37).

Keddy, P. A. 2005. Milestones in ecological thought—A canon for plant ecology. Invited Perspective. *Journal of Vegetation Science* 16:145–50.

Kellomäki, S., ed. 1998. *Forest resources and sustainable management, volume 2.* Helsinki, Finland: Fapet Oy.

Kelty, M. J., B. C. Larson, and C. D. Oliver. 1992. *The ecology and silviculture of mixed-species forests.* Dordrecht, The Netherlands: Kluwer Academic Publishers.

Kerr, G. 1999. The use of silvicultural systems to enhance the biological diversity of plantation forests in Britain. *Forestry* 72:191–205.

Kimmins, J. P. 1992. *Balancing act: Environmental issues in forestry.* Vancouver, Canada: University of British Columbia Press.

Kimmins, J. P. 2004. *Forest ecology: A foundation for sustainable forest management and environmental ethics in forestry, 3rd edition.* Upper Saddle River, NJ: Pearson Prentice Hall.

Kingsland, S. E. 1991. Defining ecology as a science. In *Foundations of ecology: Classic papers with commentaries*, ed. L. A. Real and J. H. Brown, 1–13. Chicago: The University of Chicago Press.

Kirkland, B. P., and A. J. F. Brandstrom. 1936. *Selective timber management in the Douglas-fir region.* Washington, DC: U.S. Department of Agriculture, Bulletin 1493.

Klocek, A., and G. Oesten. 1991. Optimale Umtriebszeit im Normal- und im Zielwaldmodell. *Allgemeine Forstzeitschrift* 162:92–99.

Kneeshaw, D. D., and Y. Bergeron. 1998. Canopy gap characteristics and tree replacement in the southeastern boreal forest. *Ecology* 79:783–94.

Kobe, R. K., and K. D. Coates. 1997. Models of sapling mortality as a function of growth to characterize interspecific variation in shade tolerance of eight tree species of northwestern British Columbia. *Canadian Journal of Forest Research* 27:227–36.

Kohm, K. A., and J. F. Franklin, eds. 1997. *Creating a forestry for the 21st century: The science of ecosystem management.* Washington, DC: Island Press.

Kolb, A., and M. Diekmann. 2005. Effects of life-history traits on response of plant species to forest fragmentation. *Conservation Biology* 19:929–38.

Köstler, J. 1949. *Waldbau.* Hamburg, Germany: Paul Parey.

Kranabetter, J. M., and P. Kroeger. 2001. Ectomycorrhizal mushroom response to partial cutting in a western hemlock-western redcedar forest. *Canadian Journal of Forest Research* 31:978–87.

Kuehne, C., and K. J. Puettmann. 2006. Groszraumstudien Nordamerikas—Neue Ansätze zur waldbaulichen Forschung (Large scale management experiments—New approaches to silvicultural research). *Forstarchiv* 77:102–9.

Kuhn, T. S. 1962. *The structure of scientific revolutions.* Chicago: University of Chicago Press.

Kuuluvainen, T. 1994. Gap disturbance, ground microtopography, and the regeneration dynamics of boreal conifer forests in Finland: A review. *Annales Zoologici Fennici* 31:35–52.

Kuuluvainen, T. 2002. Disturbance dynamics in boreal forests: Defining the ecological basis of restoration and management of biodiversity. *Silva Fennica* 36:5–12.

Langston, N. 1995. *Forest dreams, forest nightmares: The paradox of old growth in the inland west.* Seattle, WA: University of Washington Press.

Langton, C. 1990. Computation at the edge of chaos: Phase transitions and emergent computation. *Physica D* 42:12–37.

Langvall, O., and G. Orlander. 2001. Effects of pine shelterwoods on microclimate and frost damage to Norway spruce seedlings. *Canadian Journal of Forest Research* 31:155–64.

Larkin, P. A. 1977. An epitaph for the concept of maximum sustained yield. *Transactions of the American Fisheries Society* 106:1–11.

Lavender, D. P., R. Parish, C. M. Johnson, G. Montgomery, A. Vyse, R. A. Willis, and D. Winston, eds. 1990. *Regenerating British Columbia's forests.* Vancouver, Canada: University of British Columbia Press.

Lee, D. C., and B. E. Rieman. 1997. Population viability assessment of salmonids by using probabilistic networks. *North American Journal of Fisheries Management* 17:1144–57.

Lefsky, M. A., A. T. Hudak, W. B. Cohen, and S. A. Acker. 2005. Geographic variability in LIDAR predictions of forest stand structure in the Pacific Northwest. *Remote Sensing of Environment* 95:532–48.

Leibundgut, H. 1951. *Der Wald, eine Lebensgemeinschaft.* Zurich: Büchergilde Gutenberg.

Lélé, S., and R. B. Norgaard. 1996. Sustainability and the scientist's burden. *Conservation Biology* 10:354–65.

LePage, P., C. D. Canham, K. D. Coates, and P. Bartemucci. 2000. Seed abundance versus substrate limitation of seedling recruitment in northern temperate forests of British Columbia. *Canadian Journal of Forest Research* 30:415–27.

Levin, S. A. 1993. Science and sustainability. *Ecological Applications* 3:545–46.

Levin, S. A. 1998. Ecosystems and the biosphere as complex adaptive systems. *Ecosystems* 1:431–36.

Levin, S.A. 2000. Multiple scales and the maintenance of biodiversity. *Ecosystems* 3:498–506.

Levin, S.A. 2005. Self-organization and the emergence of complexity in ecological systems. *Bioscience* 55:1075–79.

Lieffers, V. J., C. Messier, P. J. Burton, J.-C. Ruel, and B. E. Grover. 2003. Nature-based silviculture for sustaining a variety of boreal forest values. In *Towards sustainable management of the boreal forest,* ed. P. J. Burton, C. Messier, D. W. Smith, and W. L. Adamowicz, 481–530. Ottawa, Canada: NRC Research Press.

Likens, G. E., F. H. Bormann, N. M. Johnson, D. W. Fisher, and R. S. Pierce. 1970. Effects of forest cutting and herbicide treatment on nutrient budgets in the Hubbard Brook Watershed-Ecosystem. *Ecological Monographs* 40:23–47.

Lindenmayer, D. B., and J. F. Franklin. 2002. *Conserving forest biodiversity: A comprehensive multiscaled approach.* Washington, DC: Island Press.

Lindenmayer, D. B., and J. F. Franklin. 2003. *Towards forest sustainability.* Washington, DC: Island Press.

Loehle, C., J. G. MacCracken, D. Runde, and L. Hicks. 2002. Forest management at landscape scales: Solving the problems. *Journal of Forestry* 100:25–33.

Loreau, M., S. Naeem, and P. Inchausti, eds. 2002. *Biodiversity and ecosystem functioning: synthesis and perspectives.* Oxford, UK: Oxford University Press.

Lorey, T., ed. 1888. *Handbuch der Forstwissenschaft.* Tubingen, Germany: Verlag der Laupp'schen Buchhandlung.

Lorimer, C. G. 1983. Tests of age-independent competition indices for individual trees in natural hardwood stands. *Forest Ecology and Management* 6:343–60.

Lovelock, J. 1979. *Gaia: A new look at life on earth.* Oxford, UK: Oxford University Press.

Ludwig, D., R. Hilborn, and C. Walters. 1993. Uncertainty, resource exploitation, and conservation: Lessons from history. *Science* 260:17–36.

MacArthur, R. H., J. W. MacArthur, and J. Preer. 1962. On bird species diversity. II. Prediction of bird census from habitat measurements. *The American Naturalist* 96:167–74.

MacArthur, R. H., and E. O. Wilson. 1967. *The theory of island biogeography.* Princeton, NJ: Princeton University Press.

Mainwaring, D. B., and D. A. Maguire. 2004. The effect of local stand structure on growth and growth efficiency in heterogeneous stands of ponderosa pine and lodgepole pine in central Oregon. *Canadian Journal of Forest Research* 34:2217–29.

Mandelbrot, B. B. 1977. *The fractal geometry of nature.* New York: W. H. Freeman and Co.

Mantel, K. 1990. *Wald und Forst in der Geschichte.* Alfeld-Hannover, Germany: M.&H. Schaper.

Margalef, R. 1969. Diversity and stability: A practical proposal and a model of interdependence. *Brookhaven Symposium Biology* 22:25–37.

Marshall, D. D., J. F. Bell, and J. C. Tappeiner. 1992. *Levels-of-growing-stock cooperative study in Douglas-fir: Report no. 10: The Hoskins study, 1963–83.* Portland, OR: U.S. Department of Agriculture, Forest Service, Pacific Northwest Research Station. Research Paper PNW-RP-448.

Marshall, D. D., and R. O. Curtis. 2002. *Levels-of-growing-stock cooperative study in Douglas-fir: Report No. 14-Stampede Creek, 30-year results.* Portland, OR: U.S. Department of Agriculture, Forest Service, Pacific Northwest Research Station, Research Paper PNW-RP-543.

Martin, G. L., and A. R. Ek.1984. A comparison of competition measures and growth models for predicting plantation red pine diameter and height growth. *Forest Science* 30:731–43.

Mason, B., and G. Kerr. 2004. *Transforming even-aged conifer stands to continuous cover management.* Edinburgh, UK: Forestry Commission, Information Note 40.

Matthews, J. D. 1989. *Silvicultural systems*. Oxford, UK: Oxford University Press.

May, R. M. 1974. Biological populations with non-overlapping generations: Stable points, stable cycles, and chaos. *Science* 186:645–47.

May, R. M. 1975. Patterns of species abundance and diversity. In *Ecology and evolution of communities*, ed. M. L. Cody and J. M. Diamond, 81–120. Cambridge, MA: Harvard University Press.

Mayr, H. 1909. *Waldbau auf naturgesetzlicher Grundlage*. Berlin, Germany: Paul Parey Verlag.

Mayer, H. 1984. *Waldbau*. Stuttgart, Germany: Gustav Fischer Verlag.

McCann, K. S. 2000. The diversity-stability debate. *Nature* 405:228–33.

McEvedy, C., and R. Jones. 1978. *Atlas of world population history*. Harmondsworth, NY: Penguin.

McPherson, G. R., and S. DeStefano. 2003. *Applied ecology and natural resource management*. Cambridge, UK: Cambridge University Press.

Mehrani-Mylany, H., and E. Hauk. 2004. Totholz: Auch hier deutliche Zunahmen. *BFW-Praxisinformation* 3:21–23.

Meidinger, D., and J. Pojar, eds. 1991. *Ecosystems of British Columbia*. Special Report Series No. 6. Victoria, Canada: British Columbia Ministry of Forests.

Messier, C., M.-J. Fortin, F. Schmiegelow, F. Doyon, S. G. Cumming, J. P. Kimmins, B. Seely, C. Welham, and J. Nelson. 2003. Modelling tools to assess the sustainability of forest management scenarios. In *Toward sustainable management of boreal forest: Emulating nature, minimizing impacts and supporting communities*, ed. P. J. Buron, C. Messier, D. W. Smith, and W. L. Adamovicz, 531–80. Ottawa: NRC Press.

Messier, C., and D. Kneeshaw. 1999. Thinking and acting differently for a sustainable management of the boreal forest. *Forestry Chronicle* 75:929–38.

Miina, J., and T. Pukkala. 2002. Application of ecological field theory in distance-dependent growth modeling. *Forest Ecology and Management* 161:101–7.

Mladenoff, D. J. 2004. Landis and forest landscape models. *Ecological Modelling* 180:7–19.

Moeur, M. 1993. Characterizing spatial patterns of trees using stem-mapped data. *Forest Science* 39:756–75.

Möller, A. 1923. *Der Dauerwaldgedanke*. Oberteuringen, Germany: Erich Degreif Verlag.

Monserud, R. A. 2002. Large-scale management experiments in the moist maritime forests of the Pacific Northwest. *Landscape and Urban Planning* 59:159–80.

Montagnini, F., and C. F. Jordan. 2005. *Tropical forest ecology: The basis for conservation and management*. Berlin, Germany: Springer-Verlag.

Morgenstern, E. K. 2007. The origin and early application of the principle of sustainable forest management. *Forestry Chronicle* 83:485–89.

Morosov, G. F. 1920. *Die Lehre vom Wald*. Berlin, Germany: Radebeul.

Muller, F., R. Hoffmann-Kroll, and H. Wiggering. 2000. Indicating ecosystem integrity: Theoretical concepts and environmental requirements. *Ecological Modelling* 130:13–23.

Mustian, A. P. 1976. History and philosophy of silviculture management systems in use today. In *Uneven-aged management and silviculture in the United States*, 1–17. U.S. Department of Agriculture, Forest Service, General Technical Report WO-24.

Naeem, S. 2002. Ecosystem consequences of biodiversity loss: The evolution of a paradigm. *Ecology* 83:1537–52.

Nordén, B., and H. Paltto. 2001. Wood-decay fungi in hazel wood species richness correlated to stand age and dead wood features. *Biological Conservation* 101:1–8.

Nyland, R.D. 2002. *Silviculture: Concepts and applications, 2nd edition.* New York: McGraw-Hill.

Odum, E. P. 1969. The strategy of ecosystem development. *Science* 164:262–70.

O'Hara, K. L. 1989. Stand growth efficiency in a Douglas-fir thinning trial. *Forestry* 62:409–18.

O'Hara, K. L. 2001. The silviculture of transformation: A commentary. *Forest Ecology and Management* 151:81–86.

O'Hara, K. L., E. Lähde, O. Laiho, Y. Norokorpi, and T. Saksa. 2001. Leaf area allocation as a guide to stocking control in multiaged, mixed-conifer forests in southern Finland. *Forestry* 74:171–85.

O'Hara, K. L., and L. M. Nagel. 2006. A functional comparison of productivity in even-aged and multiaged stands: A synthesis for *Pinus ponderosa. Forest Science* 52:290–303.

O'Hara, K. L., R. S. Seymour, S. D. Tesch, and J. M. Guldin. 1994. Silviculture and our changing profession: Leadership for shifting paradigms. *Journal of Forestry* 92:8–13.

Ohmann, J. L., M. J. Gregory, and T. A. Spies. 2007. Influence of environment, disturbance, and ownership on forest composition and structure of coastal Oregon, USA. *Ecological Applications* 17:18–33.

Oliver, C. D. 1981. Forest development in North America following major disturbances. *Forest Ecology and Management* 3:153–68.

Oliver, C. D., and B.C. Larson. 1996. *Forest stand dynamics.* New York: Wiley.

Otto, H.-J. 1994. *Waldökologie.* Stuttgart, Germany: Eugen Ulmer.

Palik, B. J., C. C. Kern, R. Mitchell, and S. Pecot. 2005. Using spatially variable overstory retention to restore structural and compositional complexity in pine ecosystems. In *Balancing ecosystem values: Innovating experiments for sustainable forestry*, ed. C. E. Peterson and D. A. Maguire, 285–90. Portland, OR: U.S. Department of Agriculture, Forest Service, General Technical Report PNW-GTR-635.

Palik, B. J., R. J. Mitchell, and J. K. Hiers. 2002. Modeling silviculture after natural disturbance to sustain biodiversity in the longleaf pine (*Pinus palustris*) ecosystem: balancing complexity and implementation. *Forest Ecology and Management* 155:347–56.

Palmer, M., E. Bernhardt, E. Chornesky, S. Collins, A. Dobson, C. Duke, B. Gold, R. Jacobson, S. Kingsland, R. Kranz, M. Mappin, M. L. Martinez, F. Micheli, J. Morse, M. Pace, M. Pascual, S. Palumbi, O. J. Reichman, A. Simons, A. Townsend, and M. Turner. 2004. Ecology for a crowded planet. *Science* 304:1251–52.

Parrott, L. 2002. Complexity and the limits of ecological engineering. *Transactions of the American Society of Agricultural Engineers* 45:1697–1702.

Parrott, L., and R. Rok. 2000. Incorporating complexity in ecosystem modeling. *Complexity International, Volume* 7. Paper ID: lparro01, http://www.complexity.org.au/ci/vol07/lparro01. Accessed Sept. 23, 2007.

Peng, C. 2000. Understanding the role of forest simulation models in sustainable forest management. *Environmental Impact Assessment Review* 20:481–501.

Perala, D. A. 1977. *Manager's handbook for aspen in the north-central states*. St. Paul, MN: U.S. Department of Agriculture, Forest Service, North Central Forest Experiment Station, General Technical Report NC-36.

Perera, A. H., L. J. Buse, and M. G. Weber, eds. 2004. *Emulating natural forest landscape disturbances*. New York: Columbia University Press.

Perry, D. A. 1994. *Forest ecosystems*. Baltimore, MD: Johns Hopkins University Press.

Peterken, G. F. 1996. *Natural woodland: Ecology and conservation in northern temperate regions*. Cambridge, UK: Cambridge University Press.

Peters, R. H. 1991. *A critique for ecology*. Cambridge, UK: Cambridge University Press.

Petersen, R. G. 1985. *Design and analysis of experiments*. New York: M. Dekker.

Peterson, C. E., and D. A. Maguire, eds. 2005. *Balancing ecosystem values: Innovative experiments for sustainable forestry*. Portland, OR: U.S. Department of Agriculture, Forest Service, Pacific Northwest Research Station, General Technical Report PNW-GTR-635.

Pfeil, W. 1851. *Die Forstwirthschaft nach rein praktischer Ansicht. Ein Handbuch fur Privatforstbesitzer, Verwalter und insbesondere fur Forstlehrlinge, 4th edition*. Leipzig, Germany.

Pickett, S. T. A., and P. S. White. 1985. *The ecology of natural disturbance and patch dynamics*. New York: Academic Press.

Platt, W. J. 1994. *Evolutionary models of plant population/community dynamics and conservation of southeastern pine savannas*. Proceedings of the North American Conference on Savannas and Barrens. http://www.epa.gov/ecopage/upland/oak/oak94/Proceedings/Platt.html. Accessed Apr. 27, 2008.

Platt, W. J., and D. R. Strong. 1989. Special feature: Gaps in forest ecology. *Ecology* 70:535–76.

Poage, N. J., and P. D. Anderson. 2007. *Large-scale silviculture experiments of western Oregon and Washington*. Portland, OR: U.S. Department of Agriculture, Forest Service, Pacific Northwest Research Station, General Technical Report PNW-GTR-713.

Pommerening, A., and S. T. Murphy. 2004. Review of the history, definitions and methods of continuous cover forestry with special attention to afforestation and restocking. *Forestry* 77:27–44.

Ponge, J.-F. 2005. Emergent properties from organisms to ecosystems: Towards a realistic approach. *Biological Reviews* 80:403–11.

Pretzsch, H. 2005. Diversity and productivity in forests: Evidence from long-term experimental plots. In *Forest diversity and function: Temperate and boreal systems*, ed. M. Scherer-Lorenzen, C. Körner, and E.-D. Schulze, 41–64. Heidelberg, Germany: Springer-Verlag.

Pretzsch, H., P. Biber, and J. Dursky. 2002. The single tree-based stand simulator SILVA: Construction, application and evaluation. *Forest Ecology and Management* 162:3–21.

Proulx, R. 2007. Ecological complexity for unifying ecological theory across scales: A field ecologist's perspective. *Ecological Complexity* 4:85–92.

Puettmann, K. J. 2000. Ecosystem management als neue Grundlage fur die Waldbewirtschaftung in Nordamerika (Ecosystem management as new forest management paradigm in North America). *Forstarchiv* 71:3–9.

Puettmann, K. J., and C. Ammer. 2007. Trends in North American and European regeneration research under the ecosystem management paradigm. *European Journal of Forest Research* 126:1–9.

Puettmann, K. J., and A. Ek. 1999. Status and trends of silvicultural practices in Minnesota. *Northern Journal of Applied Forestry* 16:203–10.

Radtke, P. J., J. A. Westfall, and H. E. Burkhart. 2003. Conditioning a distance-dependent competition index to indicate the onset of inter-tree competition. *Forest Ecology and Management* 175:17–30.

Rietkerk, M., J. van de Koppel, L. Kumar, F. van Langevelde, and H. Prins. 2002. The ecology of scale. *Ecological Modelling* 149:1–4.

Rivington, M., K. B. Matthews, G. Bellocchi, K. Buchan, C. O. Stockle, and M. Donatelli. 2007. An integrated assessment approach to conduct analyses of climate change impacts on whole-farm systems. *Environmental Modelling & Software* 22:202–10.

Roberts, M. 2004. Response of the herbaceous layer to natural disturbance in North American forests. *Canadian Journal of Botany* 82:1273–83.

Röhrig, E., N. Bartsch, and B. v. Lüpke. 2006. *Waldbau auf ökologischer Grundlage*. Stuttgart, Germany: UTB Verlag Eugen Ulmer.

Rothe, A., and D. Binkley. 2001. Nutritional interactions in mixed-species forests. *Canadian Journal of Forest Research* 31:1855–70.

Roughgarden, J. 1979. *Theory of population genetics and evolutionary ecology: An introduction*. New York: Macmillan.

Rubner, K., and W. Leiningen-Westerburg. 1925. *Die pflanzengeographischen Grundlagen des Waldbaus, 2nd edition*. Neudamm, Germany: J. Neumann.

Rudolf, P. O. 1985. *History of the lake states forest experiment station*. St. Paul, MN: U.S. Department of Agriculture, Forest Service, North Central Forest Experiment Station.

Ruppert, C. 2004. Der kommunale Forstbetrieb im Spannungsfeld von Gemeinwohlorientierung und Erwerbswirtschaft. Eine Analyse der Möglichkeiten von Rechts- und Organisationsformen. In *GFH-Mitteilungen* 16:9–10.

Salo, S., and O. Tahvonen. 2002. On the optimality of a normal forest with multiple land classes. *Forest Science* 48:530–42.

Sander, I. L. 1977. *Manager's handbook for oaks in the north-central states*. St. Paul, MN: U.S. Department of Agriculture, Forest Service, North Central Forest Experiment Station, General Technical Report NC-37.

Sarr, D., and K. J. Puettmann. 2008. Forest management, restoration, and designer ecosystems: Integrating strategies for a crowded planet. *Ecoscience* 15:17–26.

Sarr, D., K. J. Puettmann, R. Pabst, M. Cornett, and L. Arguello. 2004. Restoration ecology: New perspectives and opportunities for forestry. *Journal of Forestry* 102:20–24.

Schama, S. 1995. *Landscape and memory*. New York: Knopf.

Scherer-Lorenzen, M., C. Körner, and E.-D. Schulze. 2005. The functional significance of forest diversity: The starting point. In *Forest diversity and function: Temperate and boreal systems ecological studies, Vol. 176*, ed. M. Scherer-Lorenzen, C. Körner, and E.-D. Schulze. Berlin, Germany: Springer.

Scherer-Lorenzen, M., E.-D. Schulze, A. Don, J. Schumacher, and E. Weller. 2007. Exploring the functional significance of forest diversity: A new long-term experiment with temperate tree species (BIOTREE). *Perspectives in Plant Ecology, Evolution and Systematics* 9:53–70.

Schönenberger, W. 2001. Cluster afforestation for creating diverse mountain forest structures: A review. *Forest Ecology and Management* 145:121–28.

Schuetz, J. P. 2001. *Der Plenterwald und weitere Formen strukturierter und gemischter Wälder.* Berlin, Germany: Parey Verlag.

Schwilk, D. W. 2003. Flammability is a niche construction trait: Canopy architecture affects fire intensity. *American Naturalist* 162:725–33.

Seymour, R. S. 2005. Integrating disturbance parameters into conventional silvicultural systems: Experience from the Acadian forest of northeastern North America. In *Balancing ecosystem values: Innovating experiments for sustainable forestry*, ed. C. E. Peterson and D. A. Maguire, 41–48. Portland, OR: U.S. Department of Agriculture, Forest Service, Pacific Northwest Research Station, General Technical Report PNW-GTR-635.

Seymour, R. S., J. Guldin, D. Marshall, and B. Palik. 2006. Large-scale, long-term silvicultural experiments in the United States: Historical overview and contemporary examples. *Allgemeine Forst-und Jagdzeitung* 177:104–12.

Seymour, R. S., and M. L. Hunter, Jr. 1992. *New forestry in eastern spruce-fir forests: Principles and applications to Maine.* Orono, ME: Maine Agricultural Experiment Station. Misc. Publ. 716.

Seymour, R. S., and M. L. Hunter, Jr. 1999. Principles of ecological forestry. In *Maintaining biodiversity in forest ecosystems*, ed. M. L. Hunter, Jr., 22–61. Cambridge, UK: Cambridge University Press.

Seymour, R. S., and L. S. Kenefic. 2002. Influence of age on growth efficiency of Tsuga canadensis and Picea rubens trees in mixed-species, multi-aged northern conifer stands. *Canadian Journal of Forest Research* 32:2032–42.

Seymour, R. S., A. S. White, and P. G. de Maynadier. 2002. Natural disturbance regimes in northeastern North America: Evaluating silvicultural systems using natural scales and frequencies. *Forest Ecology and Management* 155:357–67.

Shi, H., and L. Zhang. 2003. Local analysis of tree competition and growth. *Forest Science* 49:938–55.

Silvertown, J. 2004. Plant coexistence and the niche. *Trends in Ecology and Evolution* 19:605–11.

Simberloff, D. 1988. The contribution of population and community biology to conservation science. *Annual Review of Ecology and Systematics* 19:473–511.

Simberloff, D. S., and E. O. Wilson. 1969. Experimental zoogeography of islands: The colonization of empty islands. *Ecology* 50:278–96.

Smith, D. M. 1970. Applied ecology and the new forest. In *Western reforestation coordination committee proceedings*, 3–7. Portland, OR: Western Forestry and Conservation Association.

Smith, D. M. 1972. The continuing evolution of silviculture practices. *Journal of Forestry* 70:89–92.

Smith, D. M., B. C. Larson, M. J. Kelty, and P. M. S. Ashton. 1997. *The practice of silviculture: Applied forest ecology, 9th edition*. New York: Wiley.

Smith, F. W., and J. N. Long. 1989. The influence of canopy architecture on stemwood production and growth efficiency of *Pinus contorta* var. *latifolia*. *Journal of Applied Ecology* 26:681–91.

Solé, R. V., and J. Bascompte. 2006. *Self-organization in complex ecosystems*. Princeton, NJ: Princeton University Press.

Sorensen, T. A. 1948. Method of establishing groups of equal amplitude in plant sociology based on similarity of species content, and its application to analyses of the vegetation on Danish commons. Det Kongelige Danske Videnskabernes Selskab. Biologiske Skrifter. Bind V. Nr. 4. 1948. I. Kommission Hos Ejnar Munksgaard. Kobenhavn.

Speidel, G. 1984. *Forstliche Betriebswirtschaftslehre*. Hamburg, Germany: Paul Parey.

Spence, J. R., W. J. A. Volney, V. J. Lieffers, M. G. Weber, S. A. Luchkow, and T. W. Vinge. 1999. The Alberta EMEND project: Recipe and cooks' argument. In *Science and practice: Sustaining the boreal forest: Proceedings of the sustainable forest management network conference*, ed. T. S. Veeman, D. W. Smith, B. G. Purdy, F. J. Salkie, and G. A. Larkin, 583–90. Edmonton, Canada: Sustainable Forest Management Network.

Spies, T. A., J. F. Franklin, and M. Klopsch. 1990. Canopy gaps in Douglas-fir forests of the Cascade Mountains. *Canadian Journal of Forest Research* 20:649–58.

Spies, T. A., B. C. McComb, R. S. H. Kennedy, M. T. McGrath, K. Olsen, and R. J. Pabst. 2007. Potential effects of forest policies on terrestrial biodiversity in a multi-ownership province. *Ecological Applications* 17:48–65.

Spurr, S. H. 1956. German silvicultural systems. *Forest Science* 2:75–80.

Spurr, S. H. 1964. *Forest ecology*. New York: Wiley.

Stadt, K. J., C. Huston, K. D. Coates, Z. Feng, M. R. T. Dale, and V. J. Lieffers. 2007. Evaluation of competition and light estimation indices for predicting diameter growth in mature boreal mixed forests. *Annals of Forest Science* 64:477–90.

Steffen, W., A. Sanderson, P. D. Tyson, J. Jäger, P. A. Madson, B. Moore, F. Oldfield, K. Richardson, H.-J. Schellnhuber, B. L. Turner, and R. J. Wasson. 2004. *Global change and the earth system: A planet under pressure*. Heidelberg, Germany: Springer-Verlag.

Stephens, P. A., S. W. Buskirk, G. D. Hayward, and C. Martinez Del Roi. 2005. Information theory and hypothesis testing: a call for pluralism. *Journal of Applied Ecology* 42:4–12.

Steventon, J. D., K. MacKenzie, and T. Mahon. 1998. Response of small mammals and birds to partial cutting vs clearcutting in northwest British Columbia. *Forestry Chronicle* 74:703–13.

Stoll, P., and J. Weiner. 2000. A neighbourhood view of interactions among individual plants. In *The geometry of ecological interactions: Simplifying spatial complexity*, ed. U. Dieckmann, R. Law, and J. A. J. Metz, 11–27. Cambridge, UK: Cambridge University Press.

Struhsaker, T. T. 1998. A biologist's perspective on the role of sustainable harvest in conservation. *Conservation Biology* 12:930–32.

Tansley, A. G. 1935. The use and abuse of vegetational concepts and terms. *Ecology* 16:284–307.

Tappeiner, J. C., D. A. Maguire, and T. B. Harrington. 2007. *Silviculture and ecology of western U.S. forests*. Corvallis, OR: Oregon State University Press.

Thirgood, J. V. 1981. *Man and the Mediterranean Forest*. New York: Academic Press.

Thomasius, H. 1999. Waldbauverfahren im Wandel: Lehren aus der Geschichte. In *Kongressbericht: 100 Jahre Deutscher Forstverein*, 249–306. Schwerin, Germany: Verlag Die Werkstatt.

Thompson, D. G., and D. G. Pitt. 2003. A review of Canadian forest vegetation management research and practice. *Annals of Forest Science* 60:559–72.

Thompson, S. K. 2002. *Sampling, 2nd edition*. New York: Wiley.

Thorsen, B. J., and F. Helles. 1998. Optimal stand management with endogenous risk of sudden destruction. *Forest Ecology and Management* 108:287–99.

Tilman, D. 1999. The ecological consequences of changes in biodiversity: A search for general principles. *Ecology* 80:1455–74.

Tilman, D. 2004. Niche tradeoffs, neutrality, and community structure: A stochastic theory of resource competition, invasion, and community assembly. *Proceedings of the National Academy of Sciences* 101:10854–61.

Tilman, D., J. Knops, D. Wedin, and P. Reich. 2002a. Experimental and observational studies of diversity, productivity, and stability. In *Functional consequences of biodiversity: Empirical progress and theoretical extensions*, ed. A. Kinzig, S. Pacala, and D. Tilman, 42–70. Princeton, NJ: Princeton University Press.

Tilman, D., J. Knops, D. Wedin, and P. Reich. 2002b. Plant diversity and composition: Effects on productivity and nutrient dynamics of experimental grasslands. In *Biodiversity and ecosystem functioning: Synthesis and perspectives*, ed. M. Loreau, S. Naeem, and P. Inchausti, 21–35. Oxford, UK: Oxford University Press.

Tilman, D., J. Knops, D. Wedin, P. Reich, R. Ritchie, and E. Siemann. 1997. The influence of functional diversity and composition on ecosystem processes. *Science* 277:1300–1302.

Tomsons, S. 2001. Western ethics and resource management: A glance at the history. *Forestry Chronicle* 77:431–37.

Toumey, J. W. 1928. *Foundations of silviculture upon an ecological basis*. New York: Wiley.

Toumey, J. W., and C. F. Korstian. 1947. *Foundations of silviculture upon an ecological basis, 3rd edition*. New York: Wiley.

Troup, R. S. 1928. *Silvicultural systems*. Oxford, UK: Oxford University Press.

Uliczka, H., P. Angelstam, G. Jansson, and A. Bro. 2004. Non-industrial private forest owners' knowledge of and attitudes towards nature conservation. *Scandinavian Journal of Forest Research* 19:274–88.

Vanha-Majamaa, I., and J. Jalonen. 2001. Green tree retention in Fennoscandian forestry. *Scandinavian Journal of Forest Research* 16:79–90.

Vanselow, K. 1963. Zur geschichlichen Entwicklung der Verjüngunsformen in Deutschland. *Forstwissenschaftliche Centralblatt* 82:9–10.

Ver Hoef, J. M. 1996. Parametric empirical Bayes methods for ecological applications. *Ecological Applications* 6:1047–55.

Volkov, I., J. R. Banavar, S. P. Hubbell, and A. Maritan. 2003. Neutral theory and relative species abundance in ecology. *Nature* 424:1035–37.

von Carlowitz, G. C. 1713. *Sylvicultura oeconomica oder hauswirtschliche Nachricht und naturmassige Anweisung zur wilden Baumzucht nebst grundlicher Darstellung/Wie zu fordest durch Gottliche Benehmen dem allenthalben und insgemein eintreffenden.* Leipzig, Germany: Grossen Holz/Mangel.

Vyse, A. 1999. Is everything all right up there? A long term interdisciplinary silvicultural systems project in a high elevation fir-spruce forest at Sicamous Creek B.C. *Forestry Chronicle* 75:467–72.

Wagner, C. 1912. *Der Blendersaumschlag und sein System.* Tübingen, Germany: Laub Verlag.

Wagner, R. G. 2005. Top 10 principles for managing competing vegetation to maximize regeneration success and long-term yields. In *The thin green line: A symposium on the state-of-the-art in reforestation proceedings*, ed. S. J. Colombo, 31–35. Sault Ste. Marie, Canada: Forest Research Information Paper 160, Ontario Ministry of Natural Resources.

Wagner, R. G., and S. J. Colombo, eds. 2001. *Regenerating the Canadian forest: Principles and practice for Ontario.* Ontario, Canada: Fitzhenry & Whiteside, Markham.

Wagner, R. G., K. M. Little, B. Richardson, and K. McNabb. 2006. The role of vegetation management for enhancing productivity of the world's forests. *Forestry* 79:57–79.

Wagner, R. G., and S. R. Radosevich. 1998. Neighborhood approach for quantifying interspecific competition in coastal Oregon forests. *Ecological Applications* 8:779–94.

Wagner, R. G., and S. R. Radosevich. 1991. Neighborhood predictors of interspecific competition in young Douglas-fir plantations. *Canadian Journal of Forest Research* 21: 821–28.

Waldrop, M. M. 1992. *Complexity: The emerging science at the edge of order and chaos.* New York: Simon & Schuster.

Walker, B. H, and J. A. Meyers. 2004. Thresholds in ecological and social-ecological systems: A developing database. *Ecology and Society* 9(2):3.

Walstad, J. D., and P. J. Kuch, eds. 1987. *Forest vegetation management for conifer production.* New York: Wiley.

Waring, R. H., and S. W. Running. 1998. *Forest ecosystems: Analysis at multiple scales.* San Diego: Academic Press.

Weaver, W. 1948. Science and complexity. *American Scientist* 36:536.

Weetman, G. F. 1996. *Are European silvicultural systems and precedents useful for British Columbia silviculture prescriptions?* Victoria, Canada: Canadian Forest Service and British Columbia Ministry of Forests, FRDA Report 239.

Weetman, G., and A. Vyse. 1990. Natural regeneration. In *Regenerating British Columbia's forests*, ed. D. P. Lavender, R. Parish, C. M. Johnson, G. Montgomery, A. Vyse, R. A. Willis, and D. Winston, 118–30. Vancouver, Canada: University of British Columbia Press.

Westveld, R. H. 1939. *Applied silviculture in the United States.* New York: Wiley.

Whittaker, R. H. 1956. Vegetation of the Great Smoky Mountains. *Ecological Monographs* 26:2–80.

Whittaker, R. H. 1967. Gradient analysis of vegetation. *Biological Review* 42:207–64.

Wiedemann, E. 1925. *Der praktische Erfolg des Kieferndauerwaldes.* Berlin, Germany.

Wilson, E. O. 1988. *Biodiversity.* Washington, DC: National Academy Press.

Wilson, D., and K. J. Puettmann. 2007. Density management and biodiversity in young Douglas-fir forests: Challenges of managing across scales. *Forest Ecology and Management* 246:123–34.

Wimberly, M., and B. B. Bare. 1996. Distance-dependent and distance-independent models of Douglas-fir and western hemlock basal area growth following silvicultural treatment. *Forest Ecology and Management* 89:1–11.

Woods, A., K. D. Coates, and A. Hamann. 2005. Is an unprecedented Dothistroma needle blight epidemic related to climate change? *BioScience* 55:761–69.

Woodwell, G. M., and F. T. Mackenzie, eds. 1995. *Biotic feedbacks in the global climatic system: Will the warming feed the warming?* New York: Oxford University Press.

Zeide, B. 2001a. Reply to letter from R. Buckman. *Journal of Forestry* 99:49.

Zeide, B. 2001b. Thinning and growth: A full turnaround. *Journal of Forestry* 99:20–25.

Index

Note: page numbers in *italics* refer to figures or tables.

About the Authors

Klaus J. Puettmann is professor of silviculture and forest ecology in the Department of Forest Science, Oregon State University (OSU), Corvallis. He completed his undergraduate training in Germany at the University of Freiburg with a diploma (1986) and received his PhD in silviculture/forest modeling from OSU in 1990.

K. David Coates is senior research silviculturist with the British Columbia Forest Service, Research Section, Smithers. He received his bachelor's of science in forestry (1979) at the University of British Columbia (UBC), obtained a MSc in silviculture from Oregon State University (1987), and earned his PhD in silviculture from the University of British Columbia (1997).

Christian Messier is professor of forest ecology in the Department of Biological Sciences, University of Québec in Montréal (UQAM), and director of the Center of Forest Studies, a large interuniversity research center in the province of Québec. He obtained a BS in forestry (1984) and MSc (1986) in forest ecology from Laval University, Québec City, Canada, and a PhD (1991) from the University of British Columbia.